Other Publications:

THE GOOD COOK
THE SEAFARERS
THE ENCYCLOPEDIA OF COLLECTIBLES
THE GREAT CITIES
WORLD WAR II
THE WORLD'S WILD PLACES
THE TIME-LIFE LIBRARY OF BOATING
HUMAN BEHAVIOR
THE ART OF SEWING
THE OLD WEST
THE EMERGENCE OF MAN
THE AMERICAN WILDERNESS
THE TIME-LIFE ENCYCLOPEDIA OF GARDENING
LIFE LIBRARY OF PHOTOGRAPHY
THIS FABULOUS CENTURY
FOODS OF THE WORLD
TIME-LIFE LIBRARY OF AMERICA
TIME-LIFE LIBRARY OF ART
GREAT AGES OF MAN
LIFE SCIENCE LIBRARY
THE LIFE HISTORY OF THE UNITED STATES
TIME READING PROGRAM
LIFE NATURE LIBRARY
LIFE WORLD LIBRARY
FAMILY LIBRARY:
    HOW THINGS WORK IN YOUR HOME
    THE TIME-LIFE BOOK OF THE FAMILY CAR
    THE TIME-LIFE FAMILY LEGAL GUIDE
    THE TIME-LIFE BOOK OF FAMILY FINANCE

HOME REPAIR
AND IMPROVEMENT

HEATING AND COOLING

BY THE EDITORS OF
TIME-LIFE BOOKS

TIME-LIFE BOOKS
ALEXANDRIA, VIRGINIA

THE CONSULTANTS: Harris Mitchell, special consultant for Canada, has worked in the field of home repair and improvement for more than two decades. He is Homes editor of *Today* magazine and author of a syndicated newspaper column, "You Wanted to Know," as well as a number of books on home improvement.

Alex Melnick Jr. is a heating and air-conditioning contractor in Kensington, Maryland. He has also trained Air Force technicians in maintaining, repairing and installing both industrial and residential heating and cooling systems.

Clifford A. Wojan, formerly Professor of Mechanical Engineering at the Polytechnic Institute of New York, is a consulting engineer in thermal engineering. In that capacity, he regularly advises industry and government on heating and cooling installations, insulation and energy conservation.

Gerard Drohan, a licensed master plumber, is a specialist in alteration and repair work. He also teaches courses in basic plumbing at the Mechanics Institute in New York City.

Roswell W. Ard, a civil engineer and professional home inspector, has designed small heating and electrical systems and consulted on energy conservation techniques. He is an instructor in maritime science at Great Lakes Maritime Academy, Northwestern Michigan College, Traverse City, Michigan.

For information about any Time-Life book, please write:
Reader Information
Time-Life Books
541 North Fairbanks Court
Chicago, Illinois 60611

Contents

A Climate Made to Order

Climate-control center. A modern thermostat—more responsive than humans in controlling heating and cooling—is installed with a screwdriver and a wire-stripping tool like the one at left. Connections are made to the base *(bottom right)*, which has selector switches. The control switch is a vial of liquid mercury attached to the coil that is visible underneath the dial: as temperature expands or contracts the coil, the vial tips so that mercury either bridges contacts to turn the system on or flows away from one contact to turn the system off.

To most people the systems that heat and cool their homes seem as mysterious as their own hearts and arteries—and as difficult and dangerous to tamper with. In fact, heating and cooling systems are fairly easy to comprehend, to control and to improve. You need not accept less than adequate heating or cooling. The goal should be a comfortable climate in every room, with minimum expenditure of energy. Without becoming a master plumber or an electrical engineer, you can use the techniques in this book to achieve this goal by installing or upgrading heating and cooling equipment.

A good place to start looking for ways to improve a central heating system is at the point where fuel—gas, oil or electricity—is turned into usable heat. Most heat sources are easy to get at and adjust. A gas burner, after all, is essentially only a pipe with holes in it; you can clean it with a stiff brush and a soft wire *(page 24)*. An oil burner is no harder to tune up than a bicycle, and by increasing its efficiency you may be able to cut your heating bills 10 per cent. Electric heaters seldom need more than an occasional vacuuming.

Once the heart of your heating plant is working properly, you can profitably examine its arteries—the ducts or pipes that carry hot air, hot water or steam around the house. The most common defect—too much heat in one place and not enough in another—may be corrected merely by adding a damper to a duct *(page 11)* or by replacing a valve in a steam or hot-water convector—the finned pipes that have largely replaced the ungainly old cast-iron radiator.

Even if the arteries stop short of where you want heat to go, you can extend a duct or a pipe to a newly finished room or install an independent heater—anything from a radiant ceiling lamp *(page 76)* to a freestanding fireplace *(pages 82-85)*.

Cooling systems are as versatile as their heating counterparts. Central air conditioning is fairly simple to add if you already have a forced-air heating system *(pages 108-117)*; with a little more work you can add central air conditioning to a house equipped with steam or hot-water heat *(pages 118-119)*. And there are also many ways to cool a house that use no refrigeration at all *(pages 94-101)*.

For maximum comfort and efficiency, effective control of a heating or a cooling system is essential. The right kind of thermostat, correctly installed, will keep the temperature where you want it; a clock-controlled thermostat *(page 18)* saves fuel by turning down the system when you do not need warmth or cooling, then turning it on just before you do. With such devices and a little know-how you can disprove at least part of Mark Twain's famous witticism: Even if no one does anything about the weather outside your house, inside you can enjoy a year-round climate of your choice.

Simple Steps to Even, Economical Heat

With a few simple tricks—not one of them calling for work on a furnace or a boiler—you can increase the efficiency of any heating system. Warm air, hot water or steam make their way through ducts or pipes to the rooms of a house. Once there, they release their heat—through a register in a warm-air system; through a radiator or a finned tube called a convector in a hot water or steam system. Working on these elements alone—the ducts, pipes, registers, radiators and convectors—you may get more heat, more evenly distributed, for fewer dollars.

Start at the points where heat enters a room. Registers, radiators and convectors ought to be clean: dust and dirt absorb heat, and can block the flow of air from a register. Vacuum vanes and fins, and remove a register face if necessary to get at small objects behind it. At the same time, check air flow from the register. It should be completely unobstructed; if you must place furniture in front of a register, route the heat around the obstacle and into the room with a plastic deflector.

Radiators and convectors permit a greater variety of improvements. You can increase the heat they throw into a room by taping or tacking a sheet of aluminum foil to the wall behind them—cut the foil to fit from the floor to the top of the fixture. If convector fins are not straight, align them with pliers. A radiator cover can block air flow: check for clear openings at the bottom to admit cool air, and at the top to release warm air—if necessary, cut small openings. Metallic paint robs a radiator of heating power: you need not remove the paint, but you should repaint the radiator with nonmetallic paint—a dull black is preferable because it radiates the most heat.

To work efficiently, hot-water radiators and convectors must be completely filled with water. At least once a year, at the beginning of the heating season, purge these fixtures of air through their bleeder valves; during the heating season, purge any individual fixture that runs cooler than normal. To keep a one-pipe steam radiator from hammering or losing heat, pitch it slightly toward its inlet valve with thin wood shims (below).

Behind the scenes, yet still inside the rooms to be heated, you can make less obvious but even more effective improvements. One of them deals with the controls of the heating system. Modern thermostats are fitted with ingenious devices called anticipators, which prevent large swings of temperature in a heated room. At a setting of 70°, a thermostat might switch the heating plant on when the temperature of a room drops to 68°. The anticipator, which contains a tiny heating coil, raises the temperature of the thermostat faster than that of the room itself; the thermostat shuts the plant down before the room reaches 70°, but the heat already in the system carries the room temperature to the thermostat setting or a little beyond it. To adjust an anticipator that permits wastefully wide swings of temperature, you must work inside the thermostat—but the adjustment itself is simple.

Finally, there are the ducts and pipes that carry heat throughout the house. Perhaps the most effective single improvement you can make in your heat-distribution system is insulating these heat conductors wherever they pass through an unheated basement, crawl space or attic. Inexpensive and easy to install, pipe and duct insulation blocks the most needless of all heat losses—the heat that never gets to a room at all.

While insulation is a one-time job, balancing a heating system may have to be done often—and always whenever you finish a new room or bring heat to a hitherto unused space. The object of balancing is not only efficiency, but comfort—the comfort of a house in which all the rooms are evenly heated or, sometimes even more important, in which individual rooms are brought to different temperatures for special purposes. In balancing, dampers or valves are individually adjusted to regulate the flow of air, water or steam through ducts or pipes. For the use of this technique in warm-air systems, see pages 10-11; for hot-water and steam systems, see pages 12-13.

Setting the tilt of a steam radiator. Check the alignment of the radiator with a carpenter's level. If it tilts away from its inlet pipe, set a block of wood under the legs at the other end, under the air vent. The wood block should be just thick enough to reverse the tilt; normally, you will be able to drive it into place with a mallet after lifting or tilting the radiator slightly.

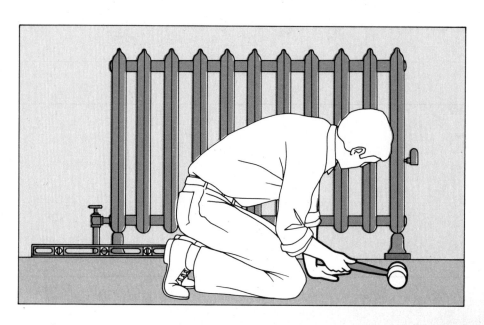

The Pros and Cons of Constant Circulation

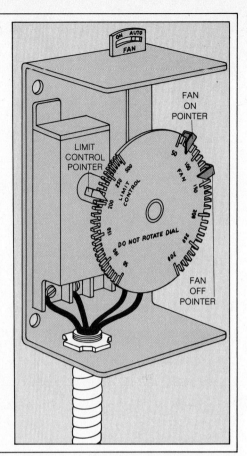

Most authorities on forced-air heating systems now favor "constant circulation"—the furnace blower runs constantly throughout the heating season. The furnace itself goes on and off, switched by the thermostat, but the blower keeps air moving through the system. Even when the furnace is off, heat remains in the system's ducts; the blower drives this heat into the rooms of a house, and the homeowner gets to use all the heat he pays for.

Along with savings in fuel costs, constant circulation makes a system heat more evenly, without frequent fluctuations in temperature. Even an individual room heats more evenly, because the drafts created when cool air moves toward the return grilles as the blower starts are largely eliminated.

But the practice has disadvantages, too. Some people argue that constant circulation is needlessly expensive—it subjects a blower to constant wear and tear, and increases the electricity bill by an amount that, depending on the relative price of different forms of energy in a specific locality, may be more than the savings in fuel.

You may want to try constant circulation in your own system for a month or two to see whether it increases your comfort and decreases your costs. To do so, check your thermostat first. If it has a "fan" lever, with settings for "on" and "automatic," set the lever at "on" for constant circulation.

If your thermostat has no fan lever, check the furnace. A control box (right) is mounted on the furnace. It may have an external switch that can be set for "on" or "automatic"; here, too, the "on" setting is the one for constant circulation. If there is no switch, remove the box cover to expose a toothed dial. Do not touch the side labeled limit control. The other side of the dial, which controls the blower, has two pointers. Set the lower ("off") pointer at about 70° and the higher ("on") pointer at about 90° to make the blower run constantly during the heating season.

Bleeding a radiator. To release trapped air, small bleeder valves are mounted near the tops of radiators and convectors. Some have a slot for a screwdriver. Others, like the one above, have a head that can be turned with a key, sold at hardware stores. Holding a cup below the spout, open the valve slowly. Air will hiss out at first; then, when all the air is discharged, water will spurt out—fast. Close the valve immediately.

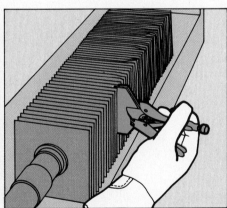

Straightening convector fins. The metal fins on convectors should provide straight, smooth paths for air rising through the fixture. If a fin is bent, twist the metal back into shape with a pair of pliers. The best tool for this job is the broad-billed pliers shown above.

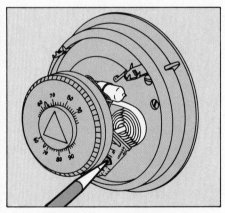

Adjusting a thermostat anticipator. Remove the thermostat cover. The anticipator, a movable pointer near the edge of the thermostat, is normally set for the amperage of the furnace. To correct wide swings of temperture above the thermostat setting, move the pointer down with the point of a pencil, .1 ampere at a time, and give the system a few hours to adjust before each change of the setting. If the furnace starts and stops too often, move the pointer to a higher setting, .1 ampere at a time.

How to Balance Your Forced-Air System

Your heating system is meant to give each room in your house the amount of heat you want in it. This does not necessarily mean keeping every room at exactly the same temperature. You may want some rooms—bathrooms or a nursery, for instance—warmer than others. You may like a bedroom on the cool side, and may send less heat to the kitchen because it already gets heat from the range and from other appliances.

Adjusting heat distribution from hot-water or steam systems is difficult; with many electric systems it is automatic, each circuit having its own thermostat; forced-air systems are in-between. Adjusting them is fairly simple if there are dampers—the movable metal plates in ducts that lead from the furnace to the outlets called registers, located in the rooms of the house. Dampers and registers have something in common—a damper regulates the flow of air through a duct, a register can regulate its flow into a room—but registers are relatively inefficient balancing tools. Their movable vanes are useful primarily to direct or diffuse the currents of warm air in a room, or to shut off the warm-air supply temporarily in a room that will be empty for an extended period.

To work with dampers, you must identify the duct that serves each room in your home; depending on the duct pattern *(below)*, the dampers themselves may be either widely scattered or clustered close to the furnace. An individual duct is wide open when the damper handle on its side parallels the duct path, it is closed when the handle is perpendicular to the duct path, and partially closed, or damped, when the handle is between these extremes. On some dampers, a locking nut will hold the damper fast in any position.

Balancing is simple, but may take a long time, spreading over several days. Balance the ducts on cold days, when the heating system is in full operation. Start by damping the duct to a room that seems too hot—preferably one that lies at the end of a relatively short duct run. (Damping at this point will send more air

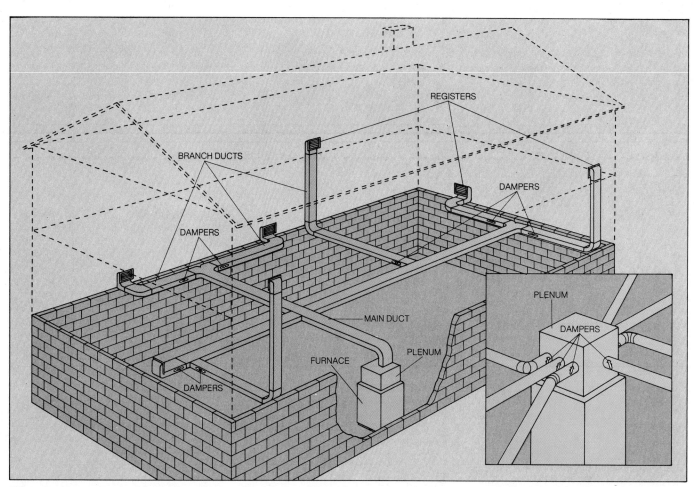

Where the dampers are. In all forced-air systems, duct runs begin at the plenum, a large chamber attached to the furnace and containing warm air under pressure. An extended-plenum system *(above)* has a main duct (or ducts) running from the plenum, and branch ducts running from a main to the registers in each room. You can usually detect the route and destination of a duct run by inspection; if necessary, close a damper to see which room turns cold. The dampers are set near the starts of the branches; to reach all of them while balancing a system, you may have to go from one end of a basement or crawl space to the other.

In a radial system *(inset)* all ducts run directly from the plenum to the registers. Dampers are near the furnace, readily accessible for balancing.

to rooms farther from the furnace.) Wait 6 to 8 hours, then check the temperature of the room, both subjectively by "feel," and objectively with a thermometer held 4 or 5 feet above the floor. Then move on to the other rooms, one at a time, partially closing dampers or opening them wide for rooms that seem too cold, until you have worked your way completely through the house.

Inevitably, you will find that your system has certain peculiarities. For example, the room containing the thermostat may heat up so quickly that the thermostat shuts the furnace down before other rooms are fully heated. The solution is to close the duct leading to the thermostat room—almost completely if necessary. More serious is the situation in which the rooms farthest from the furnace never reach a comfortable temperature even with their ducts wide open. Before you question the capacity of your furnace, try stepping up the speed of the furnace fan by the method shown on pages 37-38, to send more air to these distant rooms.

When the major balancing job is complete, you will almost certainly have to do another, for minor adjustments. At any constant fan speed, the amount of warm air forced through the house is much the same. When you increase or decrease the amount of air flowing into one room, you change the amounts flowing into the others. Go through the rooms again, adjusting one damper at a time and observing the 6-to-8-hour waiting period between each adjustment. Finally, when the system is balanced, with each room getting the amount of heat you want in it, mark the damper settings with a felt-tipped pen.

If your heating system is not fitted with dampers for this balancing job, put them in yourself—they are inexpensive and easy to install. For round duct, you may be able to buy a matching section, usually 2 feet long, that contains a factory-installed damper. Turn the furnace off and, with a hacksaw or tin snips, cut out a length of duct 4 inches shorter than the damper section at the beginning of a duct run. Install the new section by the method discussed on pages 56-57. Ready-made damper sections for rectangular ducts are harder to find because rectangular ducts come in many sizes. But you can make your own damper or have a sheet-metal shop make one for you, then install it yourself (below).

Installing a Damper in a Rectangular Duct

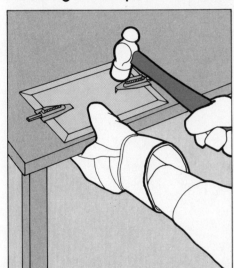

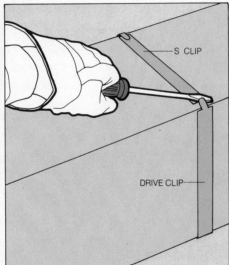

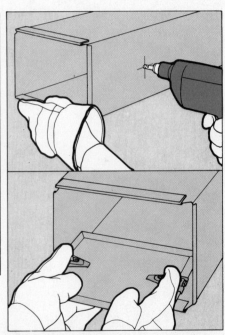

1 Making the damper. Measure the height and width of the duct. Wearing gloves use tin snips to cut a rectangle of sheet metal 1 inch longer and wider than these dimensions. Mark the dimensions of the duct within this rectangle, snip off the corners at about a 45° angle and, with a bending tool or a pair of broad-billed pliers (page 51), fold the edges of the metal back along the marked lines to form rounded edges, two layers thick.

The hardware for the damper includes two spring-loaded clips, available at hardware and heating-supply stores; buy the type that does not require welding or riveting. Slide the clips over the short edges of the damper, lining them up at the exact center. Set the damper on a firm surface and drive the clip prongs through the damper with a hammer.

2 Opening the duct. Turn the furnace off and let it cool for an hour or so. Remove the duct tape, if any, over a connection of the branch duct just past the point where the branch leaves the main duct, and open this connection. Small ducts may simply snap together and can easily be pulled apart; more often, a rectangular duct is held together by horizontal S clips and vertical drive clips. With a screwdriver, open the tabs at the tops and bottoms of the drive clips, then pull the clips down and off the duct connection with pliers. Separate the duct sections by pulling them out of the S clips.

Remove the hanger supporting the duct section that lies farther from the main duct, and carefully lower the free end of this section until it is clear; support it in this position on a convenient prop—the rung of a ladder will do.

3 Installing the damper. Mark dots on the sides of the duct at a distance from the edge equal to half the height of the damper plus 2 inches. Draw vertical lines through the dots, and horizontal lines along the centers of the sides of the duct. Where the lines intersect, drill holes the size of the bolts on the damper clips (top). Wearing a long-sleeved shirt, compress the spring-loaded bolts, slide the damper into the duct (bottom) and release the bolts into the drilled holes. Install the damper handle. To rejoin the ducts, slip their edges into the S clip. Fold the bottom tabs of the drive clips, tap these clips lightly into place with a hammer, then fold the top tabs. Cover the connection with duct tape.

S CLIP

DRIVE CLIP

Balancing Liquid Heat Systems

Valves and vents take the place of dampers in heating systems that circulate water or steam, but the methods for tuning the temperatures of individual rooms are much the same as those of a forced-air system. The text on pages 10-11 describes these methods in detail; the major difference in practice is that water and steam do not permit the precision of forced-air balancing.

In a hot-water system, flow valves near the starting point of branch lines roughly correspond to warm-air dampers, inlet valves at convectors or radiators correspond to the movable vanes of registers. Few systems include both types of valve; use the flow valves for balancing if you have them, the inlet valves if you do not.

Steam systems are less flexible: normally, their only balancing device is an adjustable air vent mounted on each radia-tor or convector. In the absence of such a vent you can make crude adjustments at the inlet valve at the other end of the radiator, but adjustable vents are inexpensive and easy to install.

At a higher price, you can substitute inlet valves with thermostatic controls to keep the system balanced in all weathers. This thermostatic valve is not set at a specific temperature but for a temperature range—65° to 70°, for example. A sensing device inside the valve regulates the flow of steam into the radiator or convector; when the temperature of the room is within the range of the thermostat setting, the inlet valve closes completely. Even if you choose to control the temperature with a valve control, however, the air vent should be cleaned periodically—and should always be cleaned as part of the balancing job.

Adjusting Hot Water Flow

Where the valves are. No hot-water system would contain as many valves as the one diagramed below, but all systems resemble this one in certain major features. A furnace heats water, a circulator pumps the hot water through main and branch supply pipes to convectors (or, in some systems, radiators), and return pipes carry cooled water back to the furnace to be reheated (variations of this basic system are shown on page 66). The diagramed system includes both balancing valves (red), installed in branch lines and opened or closed with a screwdriver (opposite, top), and inlet valves (blue), mounted on convectors and turned by hand. To identify the branch lines, close all the balancing valves on a cool day and set the thermostat at 68°. Open a balancing valve and wait about an hour while the convector it serves heats up. Tag that valve with the name of the room, then open the others, one by one, until all the valves are tagged.

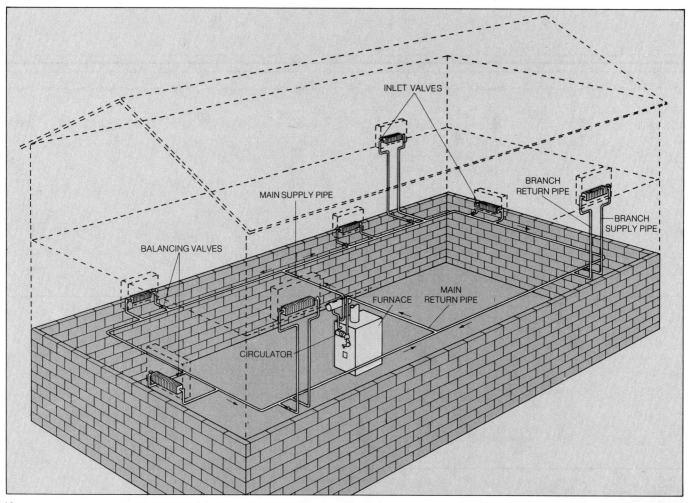

INLET VALVES

MAIN SUPPLY PIPE

BRANCH RETURN PIPE

BRANCH SUPPLY PIPE

BALANCING VALVES

MAIN RETURN PIPE

FURNACE

CIRCULATOR

Adjusting a balancing valve. The screwdriver slot that serves as a control device for this valve also indicates its setting: the valve is wide open when the slot is parallel to the path of the pipe, closed when the slot is perpendicular to the path. To regulate the flow of hot water to a convector or radiator, set the valve between these extremes and, when you have made all your balancing adjustments, mark the setting of the valve slot with a felt-tipped pen.

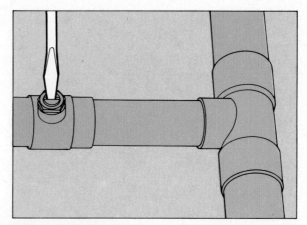

Setting the Warm-up Time in a Steam Radiator

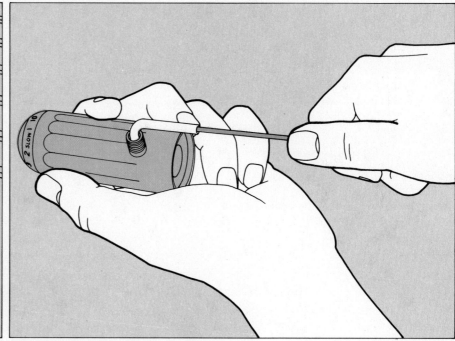

An adjustment air vent. The settings of this vent indicate exactly what you might expect—lower numbers mean less heat, higher numbers mean more. At the high-number settings, air leaves the radiator and hot steam enters it more quickly. Use the lower settings on radiators that tend to overheat a room—particularly those nearest the furnace—so that air leaves these radiators more slowly, while the others catch up. Some adjustable vents, like the one above, have a slot for a screwdriver; others have a locking nut. On the second type, loosen the nut and turn the movable cap to line up the setting number with an arrow on the vent body; then, holding the cap firmly, tighten the locking nut.

Cleaning an air vent. To keep dirt from clogging an air vent, clean the vent before balancing the system and whenever a radiator consistently runs cold. Turn off the furnace, wait an hour or so, and remove the vent. Let it soak in gasoline or benzine overnight; then rinse it off and ream the openings with a stiff wire.

The Thermostat: Key to Temperature Control

Most thermostats, whether they are the low-voltage models generally used in home heating and cooling systems or the 120- or 240-volt types that control some electric heaters *(pages 72-77)* and attic fans *(pages 94-97)*, consist of three sections. These are a cover, a middle section containing the temperature-control mechanism and a base with wire terminals and switches.

All of these—even the cover—can cause trouble. If the cover is put on incorrectly or is accidentally struck, it can jam the bimetallic coil that, by loosening or tightening as the metals expand or contract with temperature change, controls the system. Then the furnace or air conditioner may fail to start or to turn off. Often, you need only adjust the cover to get the furnace going. The base can give difficulty too. If it has slipped so that it is no longer level, the thermostat may not operate correctly. You can fix it by removing the thermostat from the base to check that the base is level *(page 16)*.

Dust and dirt are more common problems. In an old type of thermostat—one having an external mechanical contact switch attached to the bimetallic coil rather than the newer sealed mercury vial switch—you may need to clean the switch contacts *(opposite, left)*. But in any thermostat, dust and lint on the bimetallic strip can reduce its sensitivity and cause room temperature to swing between too warm and too cool; dust it with a lens brush available at camera stores. To expose as much of the bimetallic coil as possible, turn the thermostat dial as far as it will go in both directions.

Wires near the coil can become loose and switches on the base that select heating, cooling or fan operation can become corroded. A fine-tipped screwdriver and a solvent-moistened swab will solve these problems *(opposite, below)*.

None of these repairs is difficult. Almost all thermostat covers snap off and the temperature-control section is screwed to the base for easy access. However, thermostats are delicate. When working on them, use a light touch to avoid bending the bimetallic coil or damaging other sensitive components. And be sure to turn off all power beforehand.

While these simple repairs can keep any thermostat functioning for years, you may nevertheless decide to replace your old thermostat with a new one. This may be necessary if you add central air conditioning *(pages 108-117)*, but you also may wish to replace an older, metal-contact unit with a newer, mercury-switch model *(page 16)*, or to install a fuel-saving clock thermostat *(pages 17-18)*. In most cases these are simple operations that call for removing color-coded wires from the old thermostat and attaching them to color-coded terminals of the new one.

Releasing jammed parts. A thermostat cover, cocked during installation or inadvertently bumped afterward, can jam the thermostat and prevent it from turning the furnace on or off. If one of the metal support clips is bent *(inset)*, restore it to its original position. Rotate the thermostat dial to be sure that the parts move freely and that they turn the furnace on and off. If they do not—or if the thermostat no longer functions properly—replace the thermostat *(page 16)*.

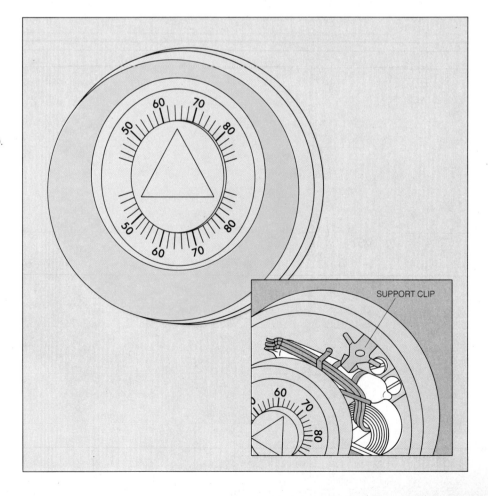

SUPPORT CLIP

Cleaning contact points. If you have a thermostat with an external contact switch attached to the coil, you may need to clean the contact points occasionally. Turn off power and remove the cover. Lower the setting to separate the points, and insert an index card between them. Never use an abrasive such as emery or sandpaper. Close the points by raising the setting. Move the card back and forth several times.

Tightening wire connections. Turn off the master switch on the heating-cooling system. Remove the cover and tighten any terminal screws that are accessible in the temperature-control section. In both the newer, mercury-switch type (illustrated below slightly rotated to show terminal screws) and the old external-contact type *(left)*, these small screws are more likely to loosen than the larger ones holding wires to the base.

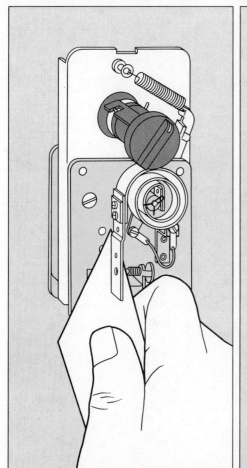

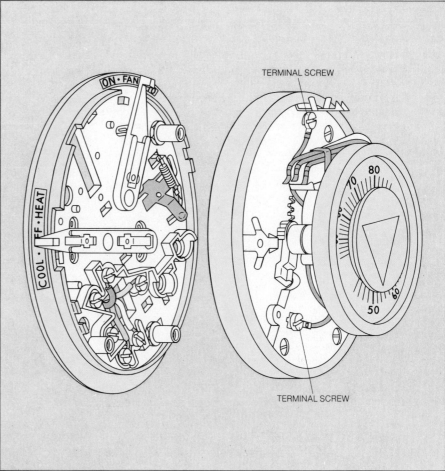

TERMINAL SCREW

TERMINAL SCREW

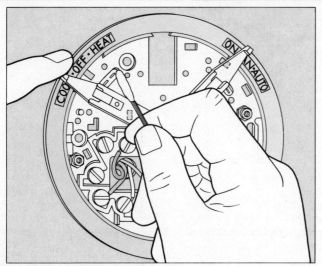

Cleaning thermostat switches. Take off the thermostat cover and, being careful not to bend the bimetallic coil, unscrew the temperature-control section from the base. Saturate a cotton swab with a nonsilicone tuner cleaner, available at electronics stores, or a strong vinegar-and-water solution, and clean the contacts near the switch levers. Move the levers from side to side to expose all the contacts. Remount the thermostat and attach the cover.

Installing New Thermostats

Replacing an old thermostat with a new one that performs the same functions at the same location involves merely connecting the new one to your existing wiring *(below, right)*. But if your replacement does something extra—automatically adjusts settings with a clock, for instance *(page 18)*—you may have to run new wiring *(opposite, above)*. And if you want to change the location of your thermostat in the course of putting in a new one, you will have to run wiring between the thermostat and furnace as you would when installing an electric heater *(pages 72-73)*. Carefully follow the manufacturers' directions for making connections.

The best place to put a thermostat is on an inside wall, out of the way of drafts and near a return vent, where the air will best reflect the average house temperature. It should be about 5 feet from the floor for convenience and away from heavily traveled areas where vibrations can jar loose the components. Keep it away from a register, a television set, a stove or a drier, direct sunshine, or an outer wall—in those locations temperatures are more extreme than anywhere else in the house, and the operation of the thermostat will be faulty.

In replacing any thermostat, make sure the new unit is compatible with the primary controls on your furnace *(page 26)* and that it is designed to carry the voltage and circuit to which it will be connected—most systems are controlled by low voltages, generally 24 volts, but some require 120 or 240 volts.

A Simple Replacement

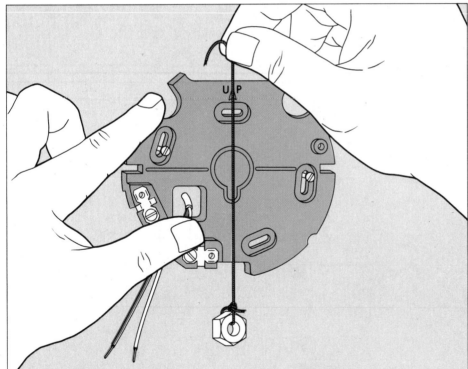

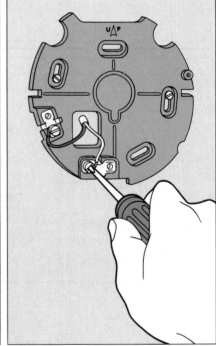

1 **Mounting the base.** Turn off all power to the heating and cooling systems. Pull off the thermostat cover. Disconnect the wires connecting the thermostat with the furnace. Depending on the kind of thermostat you have, you may be able to disconnect these wires before removing the thermostat; on some models, however, you must dismount the thermostat from the base to get at terminals on the back of the thermostat or on the face of the base. Remove the base.

Pull about 3 inches of cable through the wiring hole in the new base. Hold the base against the wall and trace the mounting holes on the wall. Plug old screw holes and the aperture around the cable to keep drafts from affecting the instrument. Attach the base, using a small spirit level or a plumb line to make sure it is level before driving the screws all the way in. Unless this type of thermostat is level, its fluid mercury temperature switch will respond inaccurately to settings.

2 **Installing the unit.** If the cable has two wires and the base two terminals, connect either wire to either terminal. If there are three wires but two terminals, connect the white wire to one terminal and the remaining wires to the other terminal. If the base has more than two terminals, match the colors of the wires to the color coding of the terminals. Check the anticipator setting *(page 9)*. Then attach the thermostat to the base and slip on the cover.

16

Regulating by the Clock

1 **Running new cable.** To install a clock thermostat of the widely used type shown on page 18, you need a 24-volt No. 18 cable containing at least four wires. Buy enough to run from the thermostat to the furnace. Turn off all power to the heating system. Disconnect and remove the thermostat and the base. Use existing wiring like fish tape *(pages 72-75)* to run the new cable from the thermostat to the furnace. By tugging at the thermostat end of the existing cable while a helper observes at the point where the cable enters the basement, determine whether the existing cable has been stapled to wall framing. If so, open an access hole in the wall *(pages 75-76, Steps 1 and 6)* and pull out the staples.

Once the existing cable is free, attach the new cable to the thermostat end of the existing cable by braiding the wires at the ends of the cables together and then wrapping the splice with electrician's tape. While a helper feeds at the thermostat end, you can use the old cable to draw the new cable through.

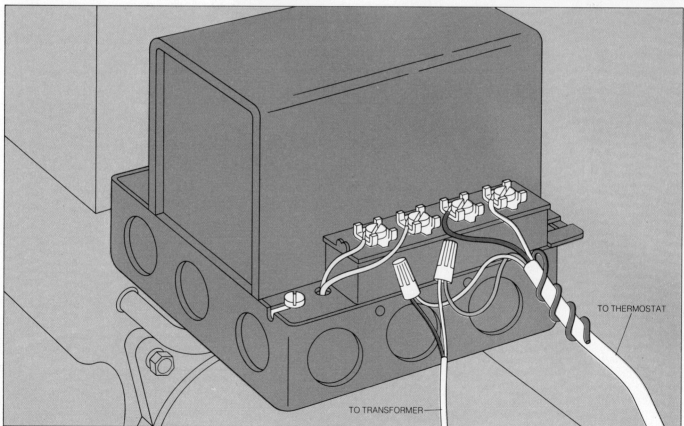

TO THERMOSTAT

TO TRANSFORMER

2 **Making connections at the furnace.** At the furnace, wind any unused (usually blue) wire of the cable from the thermostat back around the cable and strip insulation from the ends of the other wires. Attach the red and white wires to the same two furnace control terminals to which the original thermostat wires were attached. With wire caps connect the green and yellow wires to the two wires of a length of 24-volt cable (No. 18-2). Buy enough to reach from the furnace controls to a nearby junction box where you will install the clock transformer.

3 **Hooking up the clock transformer.** Turn off the power at a junction box convenient to the furnace, such as the one shown at right. Remove the light fixture from the face of the box. Attach a transformer to the box through a knockout hole. (The transformer may have a threaded nipple to which a nut can be attached or it may have a bracket-and-screw arrangement for hitching it to the box.) Remove the wire cap from the group of wires that includes the black wire of the cable from the power source. Add one of the wires from the back of the transformer to this group and refasten the wire cap.

Similarly attach the other transformer wire to the group that includes the white wire from the power source. Refasten the light fixture. Strip insulation from the two wires at the free end of the No. 18-2 cable you attached to the thermostat cable in Step 2. Attach the red wire to the transformer terminal marked R and the white wire to the C (for common) terminal.

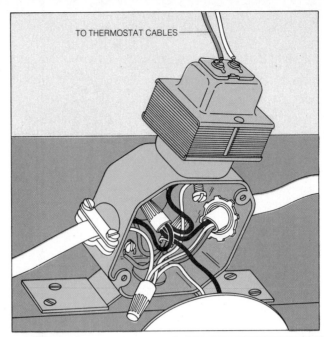

TO THERMOSTAT CABLES

4 **Wiring the base.** Level and mount the base as shown on page 16, Step 1. Wrap the unused blue wire back around the cable and strip insulation from the cable's other wires. Attach the red and white wires to the terminals marked R and W. Attach either the green or the yellow wire to either of the two clock terminals and the remaining wire to the other clock terminal.

Note whether the manufacturer's instructions direct you to make these connections with straight wire ends as shown at right. In such cases, curling the wires around the terminal screws might obstruct contact areas next to the terminals and interfere with operation of the thermostat. Plug the hole around the cable. Fit the thermostat over the mounting lugs at the top of the base and gently push the unit into place. Secure the unit with its mounting screws.

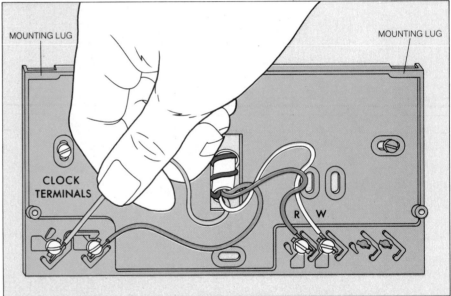

MOUNTING LUG   MOUNTING LUG

CLOCK TERMINALS

R W

5 **Adjusting the thermostat.** Restore power to the circuits involved. Set the clock, using the set wheel as shown at right. (In making adjustments, handle only those parts specified in the manufacturer's instructions.) The set wheel controls both the clock and the 24-hour timer dial. When setting the clock during the hours between 6 a.m. and 6 p.m., turn the wheel until the clock shows the correct time and the arrow above the clock points to the white side of the dial. During the period 6 p.m. to 6 a.m., turn until the arrow points to the black side.

Move the HI and LO pointers to the hours at which you want heat to increase and decrease. Set the temperature levers at the top to the high and low readings you want to maintain. Check the anticipator adjustment (*page 9*).

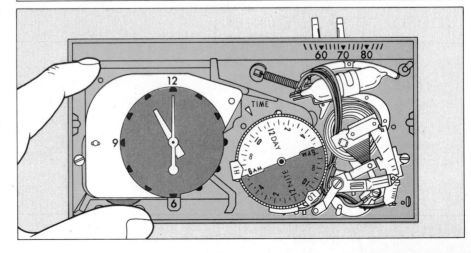

Controlling
Electric Heaters

To make the most effective use of an electric baseboard heater *(pages 76-77)* you can install a thermostat if none was provided with the heater.

The easiest method is to put in a line-voltage thermostat—one that operates on 120 or 240 volts; it gives you greater flexibility in controlling the heater than an on-off switch would, and does not require installing the relays and transformers that you would need to control the heater with a low-voltage thermostat.

Putting in a line-voltage thermostat is like installing a wall switch *(page 76, Step 5)*. Cables run from the heater and from the power source to a switch box on the wall. The thermostat is connected to both cables and attached to the switch box. The thermostat shown here is for the kind of baseboard heater most often installed—one operating on 120 to 240 volts. The wiring is appropriate for a 208- to 240-volt installation.

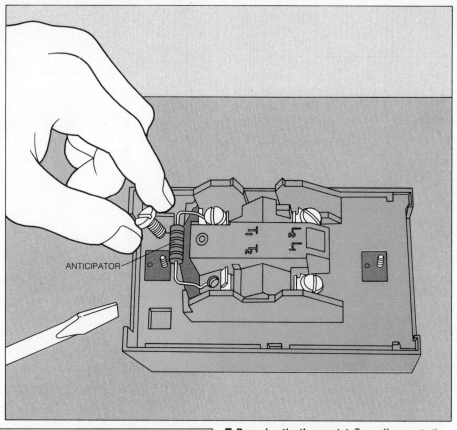

ANTICIPATOR

1 **Preparing the thermostat.** Turn off power to the switch box. Leave the cover on the thermostat and do not remove the cardboard inserts on the captive mounting screws or the insert below the anticipator. Remove the two lower terminal screws on the back of the thermostat. Insert the ends of the two small wire clamps attached to the anticipator into the screw holes as shown above. Replace the two terminal screws. (Note: If the wattage of your heater or combination of heaters totals more than 2,500, do not insert the anticipator. With such heaters, the boost the anticipator is designed to give the thermostat is not only unnecessary but will cause the thermostat to "swing," i.e., turn off and on at temperatures other than those for which it is set.)

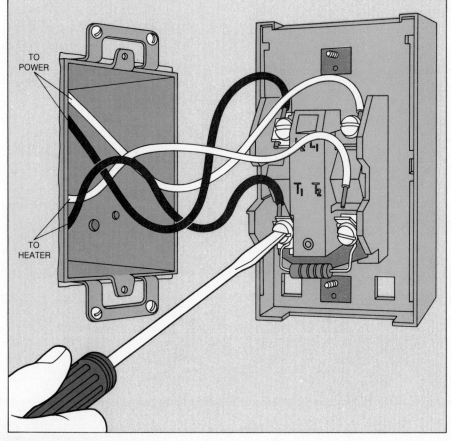

TO POWER

TO HEATER

2 **Wiring the thermostat.** Strip the ends of the wires from both the heater and the power supply. Connect the black wire from the heater to one of the terminal screws marked T on the back of the thermostat, and the black wire from the power supply to the L terminal directly above it. Attach the white wire from the heater unit to the other T terminal and attach the white power-supply wire to the other L terminal. Mark the white wires with black paint or tape to show that they are hot in this installation. Remove the thermostat cover, mount the thermostat on the box and replace the cover.

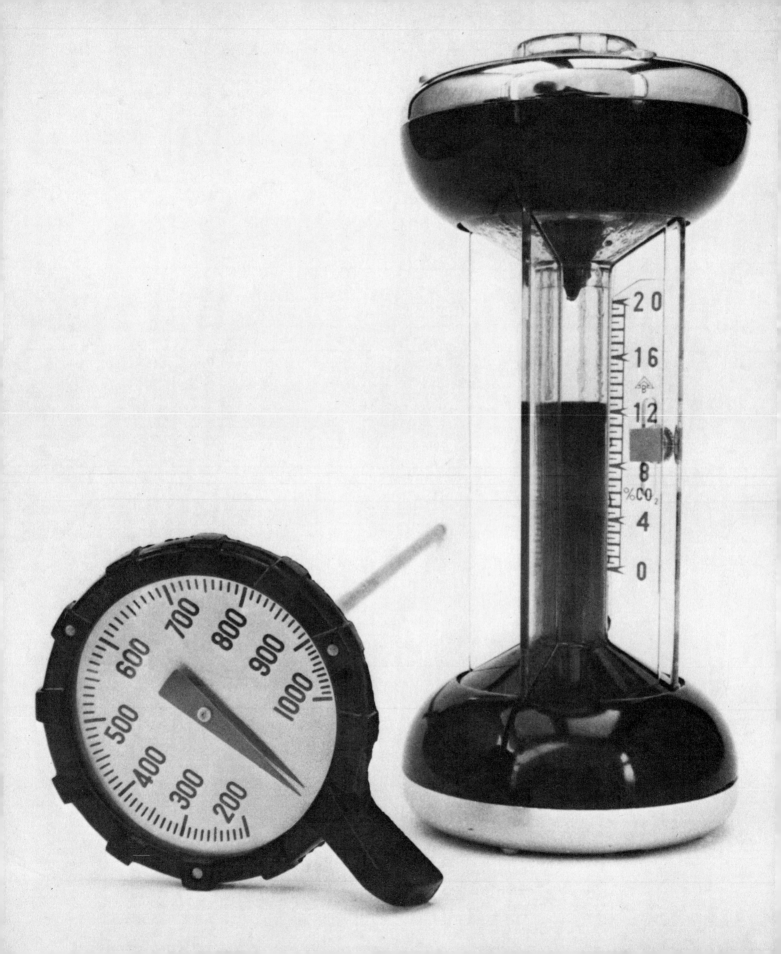

2 The Most Heat for the Least Fuel

Furnaces and their appurtenances are among the most forbidding items of household equipment. The flame is big and dangerous, the soot obnoxious and the controls arcane. Yet much furnace maintenance is simple, and even major repairs and adjustments—if approached with the correct tools—are in many cases better done by the homeowner himself than by a repairman.

Before precision-testing tools were readily available, a serviceman usually checked out a furnace by rough rules of thumb. Through the open fire door he squinted and sniffed and then adjusted the furnace's air intake to achieve an odorless yellow flame. To measure heat loss, he spat on the stack and timed the rate of evaporation. Some veteran servicemen still get passable results by squinting, sniffing and spitting, but test instruments like those at left offer a sure and easy way to make the adjustments, repairs and alterations that curb a furnace's appetite and improve its performance.

A number of adjustments can be made without special equipment. One of the simplest and most effective economy measures is a yearly clean-up, easily accomplished with a wire brush and a vacuum cleaner. The work is messy, but clearing flue passages and removing the insulating soot and scale from heat-exchanger surfaces can trim a fuel bill by as much as 10 per cent.

Cleaning and adjusting the burner itself can lead to even more significant savings. This is particularly true of an oil burner, which tends to get dirty faster than a gas one and can lose as much as 10 per cent of its efficiency after 10 weeks of operation. An afternoon spent on cleaning and tuning (pages 22-33) will reduce waste substantially, and such modifications as a new combustion chamber liner (pages 34-35) may increase burner efficiency dramatically.

Once your furnace is clean and adjusted, keeping it running efficiently is often simply a matter of attending to its regular needs. An important requirement that many people overlook is the need for a constant supply of fresh air—a demand easily met by making sure that your furnace room is always well ventilated.

Even if your system runs smoothly and efficiently, modernization increases comfort and improves the appearance of your home. If you have a hot-water system (pages 42-47), you can replace an unsightly radiator with a baseboard or an upright convector that takes up less space and provides more uniform heat. Adding a humidifier to a warm-air system (pages 39-41) will transform the dry heat of winter into a pleasant atmosphere in which you and your house plants will thrive. And since increased relative humidity may make a house feel warmer, you can then turn down the thermostat another degree or so without sacrificing comfort.

Making a Hot Flame for Maximum Efficiency

The key to the efficient operation of any gas or oil heating system is the burner. In a gas furnace, the burner has the simple task of mixing one gas with another—natural gas with air—in the correct proportions for complete combustion. With no moving parts, a gas burner is much less complex than an oil burner, which has a pump and a fine nozzle to atomize liquid fuel oil so it can be mixed with air for burning *(page 30)*.

Because of this complexity—and because oil has impurities that do not burn during combustion—oil burners need a yearly tune-up for efficient operation through the winter *(pages 26-29)*.

Gas burners, on the other hand, are less demanding, often operating for years without trouble. From time to time you may have to relight the pilot if it has blown out or if you have turned it off to conserve gas during the summer. Most furnaces have pilot instructions, which are basically the same as those below, near the control valve. Never light the pilot or make any furnace repair if there is a strong odor of gas; call the emergency service unit of your gas company. Furthermore, never try to light the pilot more than twice. If it fails to light the second time, replace the thermocouple *(opposite)*—a device that turns off gas to the furnace to prevent explosions—before trying again, or call a serviceman.

After a gas furnace has been in service for several years, rust and scale build-up on interior surfaces of the furnace can flake off and fall into the burner area. Such residues can clog burner orifices, reducing the heat output of the furnace. The remedy is to unplug the openings and give the furnace a thorough cleaning with a wire brush and vacuum cleaner. The cleaning techniques shown overleaf for a gas forced-air furnace also apply to oil-fired forced-air furnaces, as well as to boilers heated by either fuel. The spaces inside boilers, however, are generally less accessible; a smaller wire brush or a piece of clothes hanger may be needed to poke out debris before vacuuming.

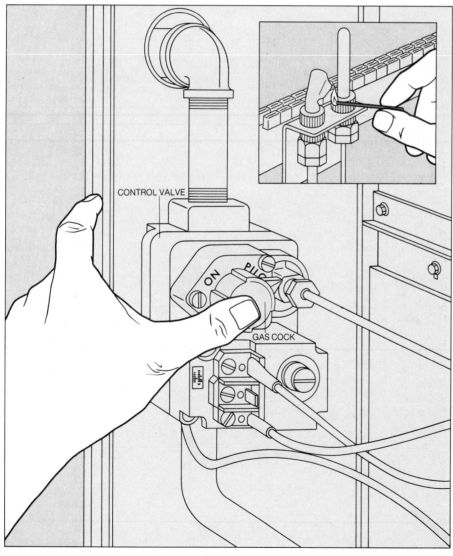

CONTROL VALVE

ON   PILOT

GAS COCK

Lighting the pilot. Remove the access panel from the front of the furnace. Examine the control valve. If it is a combination valve, it has a gas cock with a "pilot" setting *(left)*; be sure the manual shutoff valve is on *(opposite)*. If there is only a reset button on the control valve, turn off the manual shutoff valve and open the smaller pilot valve nearby. Check the service panel and emergency switch to be sure the furnace has electrical power, then turn the gas cock on the combination valve to "pilot." Depress the gas cock (or reset button) and light the pilot *(inset)*. Hold down the gas cock or button for 30 seconds, then release it. Turn on the gas to the burners by rotating the gas cock to "on" or opening the manual valve. Check the pilot flame to be sure it envelops at least half an inch of the thermocouple *(page 24)*; if it does not, adjust the pilot *(opposite)*.

If the pilot fails to light, or to stay lit, wait five minutes for gas to dissipate. Then try again, depressing the gas cock or reset button for a full minute. If the pilot goes out again, replace the thermocouple or call a serviceman.

Adjusting the pilot flame. A screw in the combination valve regulates the height of the pilot flame. On some valves, the adjustment screw is recessed and covered by a cap screw that you must remove; on others the adjustment screw is in plain view on the surface of the valve. Turn the screw counterclockwise to raise the flame, clockwise to lower it so that the thermocouple tip is well within the flame.

Adjusting the burner flame. Set the house thermostat at its highest temperature to start the furnace and keep it running. Then loosen the lock screw on an air shutter *(below)*. With your fingers, rotate the shutter open slowly until the blue base of the flame appears to lift slightly from the burner surface. Then close the shutter until the flame reseats itself on the surface, and tighten the lock screw. The flame should be blue and erect, with only occasional streaks of orange (due to impurities in the gas). Repeat this procedure for the remaining burners. Return the thermostat to its normal setting.

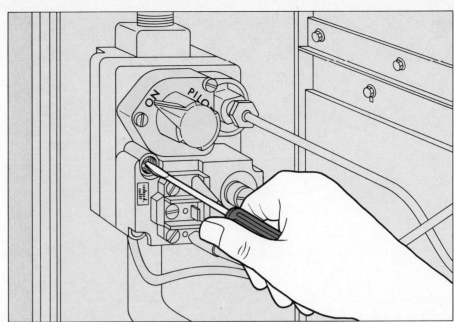

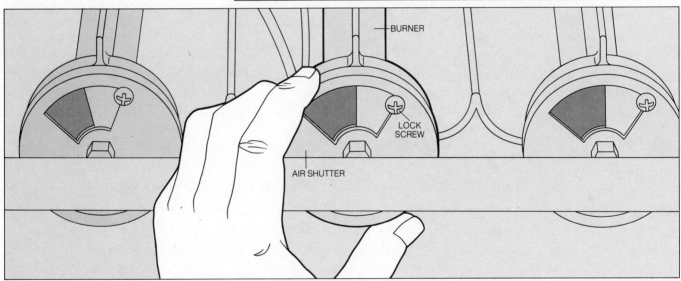

Replacing a Thermocouple

1 **Closing down the furnace.** Turn off the gas to the furnace by closing the manual shutoff valve. The valve is closed when the handle is at right angles to the gas line *(left)*. If there is a separate gas supply for the pilot, be sure to also close the manual valve on this line. Turn off the electricity by removing the furnace fuse or turning off its circuit breaker. As an added precaution, shut off the unit's main electric switch.

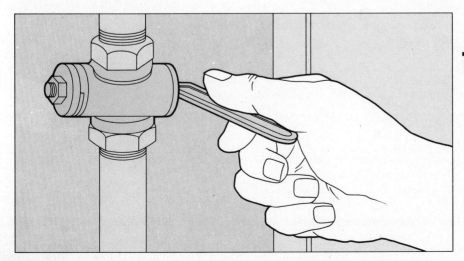

2 **Detaching the thermocouple.** After turning off the supply of gas and electricity to the furnace, wait half an hour after the pilot goes out to be sure the thermocouple is cool. Then unscrew the nut holding the thermocouple tube to the control valve *(below, left)*. Unscrew the nut that holds the thermocouple and tube to a bracket next to the pilot light *(below, right)*. Connect a new thermocouple and tube to the bracket. Then clean the control valve's threaded connection with a rag and fasten the other end of the tubing to it. Be careful not to crimp the tube if you must bend it slightly to fit. Tighten the connection nut gently with a wrench.

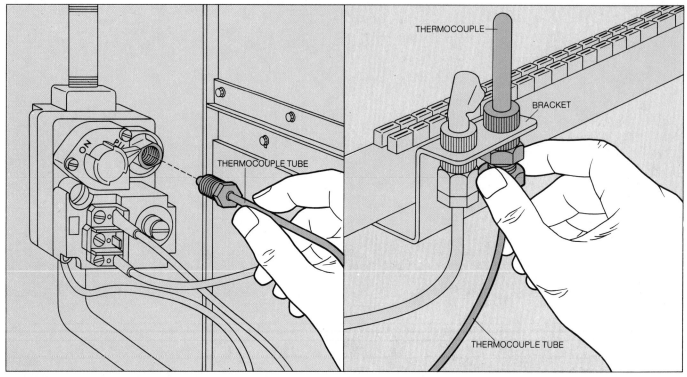

THERMOCOUPLE TUBE

THERMOCOUPLE

BRACKET

THERMOCOUPLE TUBE

Cleaning a Gas Furnace

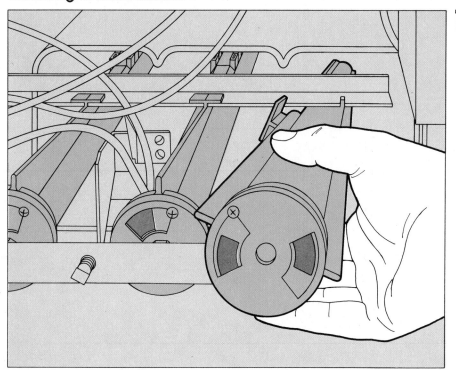

1 **Cleaning the burners.** Close down the furnace *(page 23)*, then remove the access panel covering the burners. If the pilot and thermocouple are mounted on a burner, remove them by unscrewing the bracket. In most gas furnaces, the burners are removed one at a time by first sliding them forward off the spuds, which supply gas to the burners, and then twisting and lifting the burners outward *(left)*. Some furnaces, on the other hand, have removable panels in the rear for taking out the burners.

Use a stiff brush to clean rust and dirt from burner surfaces and a soft wire—copper, for example—to unclog any plugged ports without chipping or enlarging them.

2 **Cleaning the spuds.** Run a soft wire through each of the spud openings to clear them of soot and dirt. Be careful not to damage or enlarge the openings. Spud openings are relatively large and easy to clean, but if a spud is so badly plugged that it cannot be cleaned in place, use a wrench to unscrew it from the manifold.

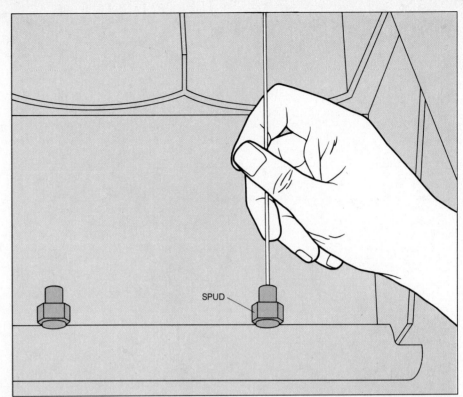

SPUD

3 **Vacuuming the furnace.** With the burners removed, use a wire brush to scrape scale, rust and soot from accessible surfaces of the combustion chamber. Unscrew the draft-diverter panel at the top of the furnace to expose the top of the heat exchanger (*below*). Remove the inserts inside the heat exchanger by pulling them out with your fingers (*inset*); some may be held in position with screws. Brush the inserts and as much of the exchanger surfaces as you can reach, then clean them with a vacuum cleaner. Replace the inserts, then vacuum the combustion chamber walls and the area below the burners. Finally, replace the draft-diverter panel.

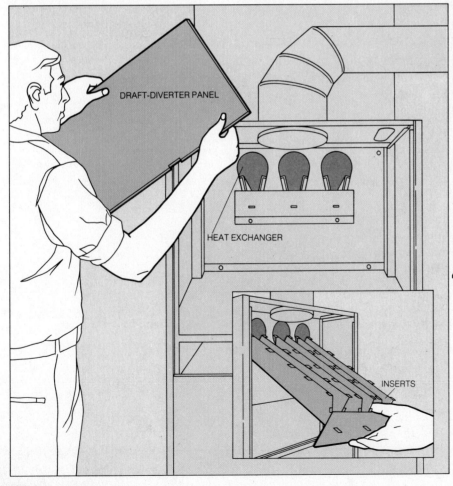

DRAFT-DIVERTER PANEL

HEAT EXCHANGER

INSERTS

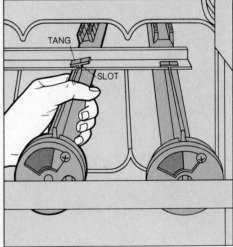

TANG

SLOT

4 **Restarting the furnace.** Reinstall the burners, sliding them into the combustion chamber and onto the spuds, or directly onto the spuds from the back of the furnace. Some burners are supported by a tang-and-slot arrangement (*above*); others rest on the floor of the combustion chamber. Reconnect the pilot and thermocouple if you removed them earlier. Turn on the gas and electricity and ignite the pilot (*page 22*). Turn up the thermostat to ignite the burner, and adjust the burner flame (*page 23*) before replacing the access panel.

Tuning Up an Oil Burner

A morning's work on your oil burner before the heating season starts can trim fuel bills, extend the burner's life and cut down on costly repairs and service calls. The process of tuning up the oil burner consists of first cleaning it thoroughly from the filter on the oil line to the nozzle on the firing assembly, then adjusting it for maximum combustion efficiency. You can undertake both of these chores safely because neither of them should involve adjusting the pump, the motor or the electric system.

Cleaning a burner calls more for patience than special skill. The work will take several hours and it is messy. You will want newspapers to protect the floor, an old pan to catch drips and a bucket to dispose of sludge and excess oil. You may also want to spread sand or cat litter in the pan to absorb oil and prevent splashing. Almost all burner parts can be cleaned with ordinary household brushes, rags and a carbon solvent such as kerosene. For parts that you need to replace such as gaskets or the filter cartridge in the oil line *(right)* be sure to get new parts of the same make and design from your oil dealer or from a heating and refrigeration supplier.

Checking the efficiency of the burner and adjusting the combustion *(pages 30-33)* is a trickier operation and requires the use of special instruments that may seem expensive. But the results you get can help to reduce the oil consumption enough so the instruments may pay for themselves in a few years.

If you do not want to invest in the instruments, you should have a serviceman make the tests for you. Instrument testing is the only way to safely adjust a burner for maximum efficiency. You can tell if the tests have been made on your furnace by looking for a ¼-inch hole in the flue *(page 30)*.

Before you do any work on a burner always shut it down completely. Turn off the master switch—or switches, if you have one at the burner and another on a wall or beside the basement stairs. Also make sure to shut off the power in the circuit that governs the burner, by switching off the circuit breaker or pulling the fuse. Then shut down the oil line at the valve between the filter and the

storage tank. If your oil line is equipped with a special fire safety valve at the pump, turn the handle clockwise to push the stem down. When the handle slips completely off the stem, give the stem a light tap with a wrench to make sure that the valve is completely closed.

Before you restore the power to the burner you must be sure to open the oil line first, to prevent the motor from drawing on an empty line and creating air pockets in the oil supply.

Cleaning the Filter System

1 Changing the oil filter. After switching off the burner and shutting off the circuit and the oil line, set a bucket or pan under the oil filter. Then loosen the bolt on the cover above the filter bowl, pull down the bowl and upend it into a bucket. The gasket around the top of the bowl and the filter cartridge inside should fall out; if the gasket sticks to the cover, pry it loose. Wipe the bowl, put in a new cartridge and gasket, and bolt the bowl back into place.

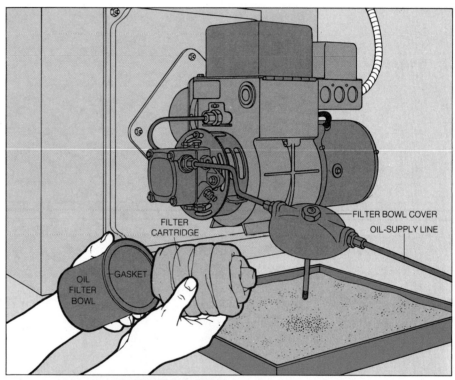

FILTER CARTRIDGE · OIL FILTER BOWL · GASKET · FILTER BOWL COVER · OIL-SUPPLY LINE

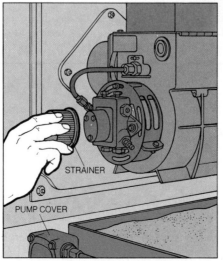

STRAINER · PUMP COVER

2 Cleaning the pump strainer. Check with your oil dealer to determine whether your pump has a strainer or a rotary-blade filter. If it has the filter, skip this step. If it has a strainer, unbolt the cover of the pump housing and lift off the cover. Discard the thin gasket around the rim of the pump. Remove the cylindrical wire-mesh strainer and soak it in solvent for a few minutes to loosen the sludge build-up. Then clean the strainer gently with an old toothbrush. If the strainer is torn or bent, get a new one. After reinserting the strainer, place a new gasket on the pump rim and bolt the cover back on.

Checking the Fan and Motor

1 Cleaning the fan. Use a long, narrow brush such as a percolator brush to sweep out the air-intake vents on the fan housing. Expose the fan by unscrewing the transformer from the top of the burner and swinging it back out of the way. Clean the fan blades with the brush and wipe the interior of the fan housing with a rag.

To reach the fan on an old-style burner *(inset)*, you must pull back the perforated bulk air band that surrounds the housing. First mark on the band and the fan housing the position of the band, then loosen the screw holding the band and pull the band back. After cleaning the fan, use the marks to set the band in its original location before you tighten the screw.

Dust on the fan and in the air-intake openings can drastically reduce your burner's efficiency, so clean them out every month or so.

2 Lubricating the motor. If your burner is equipped with small oil cups at each end of the motor, lift the lids or plugs from the cups. Then dribble four or five drops of 10- to 20-weight, nondetergent electric-motor oil in each cup and replace the lids or plugs. Lubricate the motor every two months, or at the intervals that are specified in the manufacturer's instructions. If there are no oil cups visible, the motor is self-lubricating and it is not necessary to oil it.

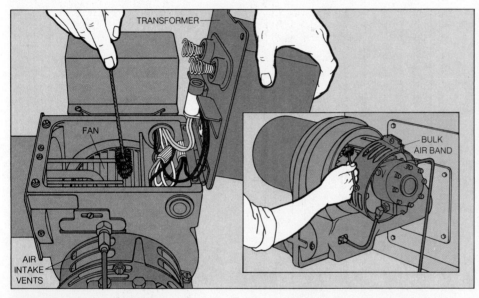

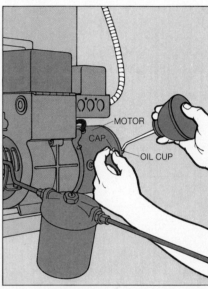

Cleaning the Sensor

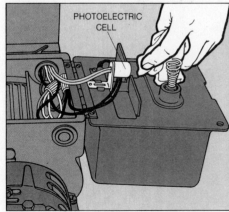

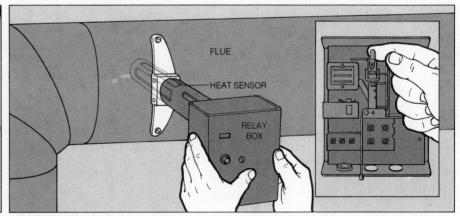

The light-detecting sensor. If the burner has a photoelectric cell that acts as a safety valve in the relay system and shuts off the motor when the ignition fails, the cell will be mounted at the end of the air tube—on the underside of the transformer or attached to the housing. Simply wipe the dirt off the cell with a clean rag.

The heat-detecting sensor. If the burner has a safety device that responds to heat to shut off the motor when the burner fails, it will be located behind the relay box on the flue. Mark the sensor tube where it meets the mounting flange. Remove the setscrew that holds the relay box to the mounting flange and pull out the box. Gently sweep the soot off the sensor with a soft-bristled paintbrush. Carefully replace the box and setscrew.

To ensure that the sensor is in contact with the shutoff mechanism, carefully remove the cover of the box. Pull the drive shaft lever toward you about ¼ inch *(inset)*, then slowly release it. Do not handle other parts of the mechanism.

Cleaning the
Firing Assembly

1 Removing the firing assembly. Mark the position of the firing assembly in the air tube. Using an open-end wrench, loosen first the flare nut and then the lock nut at the junction between the pump oil line and the nozzle oil line. Separate the two lines. Pull the entire firing assembly—the electrodes and the nozzle oil line—out of the air tube. You may need to twist the assembly as you pull it, but be careful not to knock the electrodes or nozzle against the burner housing.

To get at the firing assembly of an old-style burner (inset), take out the bolts holding the rear plate and pull the plate off. Reach into the air tube and disconnect the electrode extension rods or cables from the transformer terminals with your fingers. Mark the position of the nozzle oil line and disconnect it.

With the firing assembly removed, clean the air tube with a cloth or a brush. If there is a flame-retention device—a circular metal piece with fins or vanes—at the end of the tube, clean it also.

2 Cleaning the ignition system. Wipe soot off the electrodes and their insulators with a cloth dipped in solvent. Wipe off the electrode extension rods or cables and the transformer terminals; the contact points between the rods or cables and the terminals must be absolutely clean. If the insulators are cracked or the cables frayed, take the entire assembly to a professional for repair. Examine and clean the electrodes. The electrode tips should be spaced exactly as specified in the manufacturer's instructions—usually about ⅛ inch apart pointed toward each other, and no more than ½ inch above the center of the nozzle tip and no more than ⅛ inch beyond the front of the nozzle (inset). If necessary, loosen the screw on the electrode holder and gently move the electrodes into place.

3 Removing the nozzle. Use a pair of wrenches to grip both the nozzle body and the hexagonal adaptor at the end of the nozzle oil line. Unscrew the nozzle carefully, taking care not to twist the oil line or alter the positions of the electrodes. Examine the tip of the nozzle (inset); the stamped specifications show the firing rate in gallons of oil per hour (gph), and the angle of spray, in degrees. The type of spray pattern is usually identified by a letter of the alphabet. If the nozzle has a firing rate of 1.50 gph or less, replace it with an identical nozzle. If the nozzle has a gph of more than 1.50, you can clean and reuse it (Steps 4 and 5).

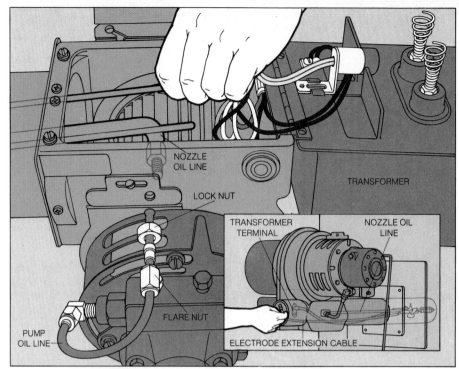

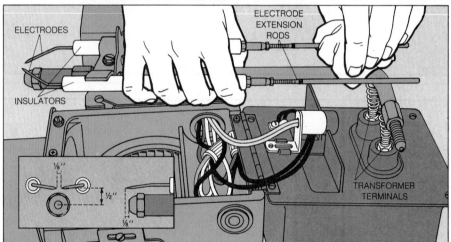

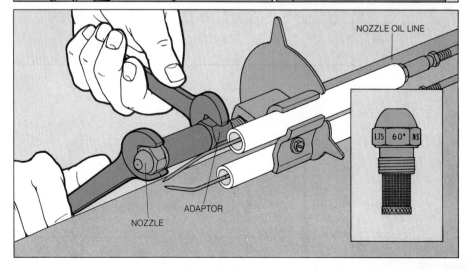

4 **Dismantling the nozzle.** With your fingers, gently unscrew the wire mesh or porous bronze strainer from the back of the nozzle. Then use a screwdriver to remove the lock nut that holds the distributor inside the tip of the nozzle. (On some nozzles, the lock nut and distributor may be a single piece.) Slide the lock nut and the distributor out of nozzle body.

5 **Cleaning the nozzle.** Soak all the nozzle parts in solvent for a few minutes, then scrub them gently with a small toothbrush. Clean the slots in the distributor with a piece of stiff paper. Clear the nozzle orifice with compressed air or a clean bristle; never use a pin or wire that might scratch the nozzle and alter the angle or pattern of its spray. Finally, flush all of the parts under

hot running water, shake them, and allow them to air-dry on a clean surface.

Reassemble the nozzle on a clean surface with clean hands and tools. Screw the nozzle back onto the adaptor with your fingers, then tighten it with the wrenches. Do not force the parts together or use the wrenches too vigorously.

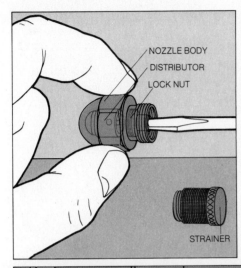

NOZZLE BODY
DISTRIBUTOR
LOCK NUT

STRAINER

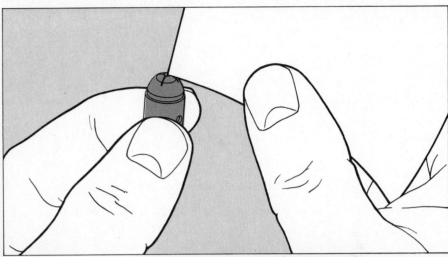

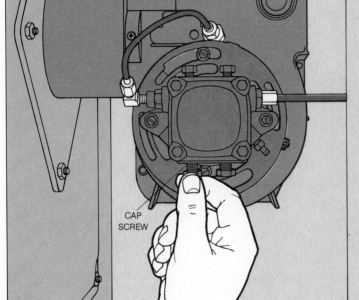

CAP
SCREW

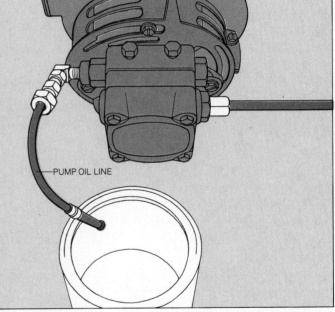

PUMP OIL LINE

6 **Priming the pump.** Loosen the cap screw on the unused intake port of the pump opposite the port to which the oil line is attached (*above, left*). Open the oil line. When oil begins to flow, allow it to run into a pan for about 15 seconds before tightening the cap screw. (If you have an underground storage tank, the oil will not feed into the pump automatically so skip this step.)

Next, loosen the nut holding the pump oil line and feed the open end of the line into a bucket (*right*). As a safety precaution, swing the transformer down or, on an old-style burner,

screw on the rear plate. Set the house thermostat several degrees above room temperature and restore power to the burner. Then station a helper at the burner master switch. When the helper throws the master switch, oil will gush out of the line with great force. Let the pump run for about 10 seconds, then turn off the motor and pull the fuse or trip the circuit breaker again.

Use the marks made in Step 1 to put the firing assembly back in the air tube, with the nozzle oil line centered in the tube. Connect the pump and nozzle oil lines by tightening the lock nut and

hexagonal nut with your fingers, then taking a quarter turn with a wrench. Screw the transformer down or, on an old-style burner, reconnect the electrodes and replace the rear plate.

When you start the burner for the winter, some air may remain in the oil line. To prevent it from building pressure in the combustion chamber, partially open the chamber's observation door and turn the burner on at the master switch. Let the burner run for 10 seconds and shut it off. Repeat five times, or until the burner shuts down smoothly and instantaneously.

How to Use Test Instruments

To make your oil burner produce the most heat with the least fuel, you need to get maximum combustion efficiency by regulating the amount of air that enters the system to burn the oil. You must also regulate the intensity of the draft that draws heat through the heat exchanger and carries away the exhaust. Too much burner air wastes heat; too little air wastes oil by failing to burn it completely—and the unburned fuel coats the heat exchanger and flues with soot. With too much draft, heat passes the exchangers before it can be absorbed; with too little draft, the combustion gases can escape through the furnace openings.

Adjusting the burner air or the draft is simple in most cases. You control burner air by adjusting the shutter at the fan (page 32) and you increase or decrease the draft with the regulator mounted on the flue (right, below). Before you tamper with the shutter or regulator, however, you must first test the burner with the instruments shown here, which are sold by heating- and refrigeration-supply dealers. (Designs vary, so read the manufacturer's instructions carefully.)

All of the tests require taking samples from the flue, so first punch a ¼-inch hole (which you can plug later with a sheet-metal screw) in the flue about a foot from the furnace, but at least 6 inches in front of the draft regulator and the flue heat exchanger (pages 60-61) if you have one. Then set the thermostat above room temperature and let the burner warm up for about 20 minutes.

Keeping the burner running continuously, test the draft in both the combustion chamber and the flue with the draft gauge. The correct combustion-chamber (overfire) draft is essential to maintain the fire, so adjust the draft before proceeding to the other tests. (Some units may not have a draft regulator. If the overfire draft readings are too high, you should have one installed.)

Complete the tests by taking smoke, carbon-dioxide and temperature readings as shown at right. With the results you can adjust the burner air and draft with precision, and pinpoint other troubles that may affect burner efficiency.

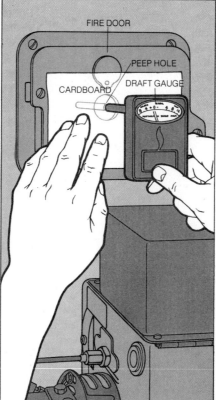

FIRE DOOR
PEEP HOLE
CARDBOARD
DRAFT GAUGE

Regulating the Draft

1 **Taking an overfire draft reading.** Punch a ¼-inch hole in a piece of thick cardboard, open the peephole in the fire door and align the hole in the cardboard with the peephole. Covering the small zero test hole at the back of the draft gauge with your finger, insert the testing tube into the combustion chamber through the hole in the cardboard. Level the gauge until the tube is horizontal and the indicator is on zero. Remove your finger from the zero test hole without disturbing the gauge. The indicator will swing to the overfire draft reading. Pull the tube out of the combustion chamber.

The correct overfire draft reading for your burner may be on the unit or listed in the manufacturer's instructions. In most cases, the reading should be between −.01 and −.02.

2 **Taking a stack draft reading.** Cover the zero test hole on the gauge with your finger and insert the tube into the ¼-inch test hole that you punched in the flue. Level the gauge until the tube is horizontal and the indicator is on zero. Remove your finger from the zero test hole. The indicator will swing to the stack draft reading. The reading can vary from −.02 to −.05.

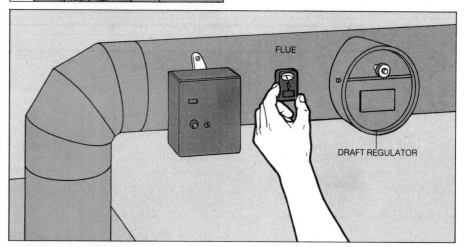

FLUE
DRAFT REGULATOR

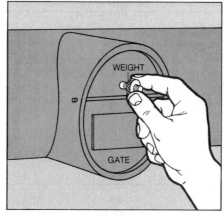

WEIGHT
GATE

3 **Adjusting the draft.** To increase the overfire draft, loosen the weight on the hinged gate of the draft regulator. Move the weight a fraction of an inch to shut the gate slightly and cut down on the amount of cool room air entering the stack. Tighten the weight and retest the draft. To decrease the draft, open the gate slightly and allow more cold air to enter the stack. It may be necessary to shut or open the gate several times to achieve the desired draft reading.

Gauging Smoke Density

1 Taking the smoke sample. Loosen the screw clamp at the end of the smoke tester, insert a strip of filter paper (supplied with the tester) in the slot and tighten the clamp. With the burner running, insert the sampling tube at least 2½ inches into the flue test hole. Slowly pull the pump handle 10 full strokes, pausing for several seconds at the end of each stroke. Then pull the tube out of the test hole, loosen the clamp and remove the filter paper.

2 Reading the smoke sample. Compare the small gray spot on the filter paper with the samples on the 10-spot smoke scale provided with the smoke tester. See the chart on page 33 to evaluate the smoke reading.

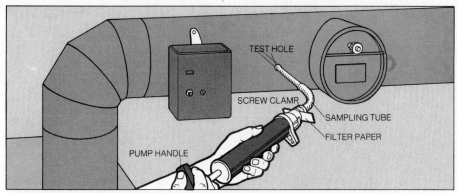

TEST HOLE

SCREW CLAMR

SAMPLING TUBE

FILTER PAPER

PUMP HANDLE

SMOKE TESTER

Taking the Readings for Combustion Efficiency

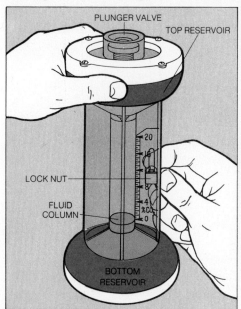

PLUNGER VALVE

TOP RESERVOIR

LOCK NUT

FLUID COLUMN

BOTTOM RESERVOIR

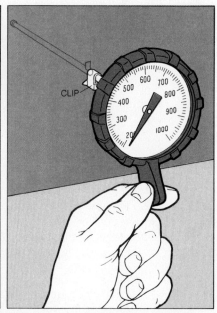

RUBBER CAP

SAMPLING TUBE

CLIP

500 600 700 800 900 1000 400 300 200

1 Preparing the carbon-dioxide indicator. Wet the inside of the indicator by inverting it and allowing all of the fluid to drain into the top reservoir. Handle the indicator carefully; the fluid inside it is corrosive. Turn the indicator upright until the fluid drains back into the bottom reservoir, then tip the indicator to a 45° angle and slowly count to five. Holding the indicator upright, depress the plunger valve to let air into the indicator. The fluid level will drop. If the fluid level is below ⅛ inch from the bottom of the central bore, dribble one drop of water at a time into the bore through the plunger valve. Loosen the lock nut on the calibrated scale and adjust it so the zero per cent mark aligns with the top of the fluid column.

2 Taking a carbon-dioxide reading. Insert the metal end of the sampling tube into the test hole. Place the rubber cap on the opposite end of the tube over the plunger valve. Hold the rubber cap firmly in place, and depress the valve with one finger. With your other hand, squeeze the rubber bulb of the tube 18 times in rapid succession. On the count of 18, release the plunger valve at the same time that you squeeze the bulb. Invert the indicator two times, letting it drain completely from one reservoir to the other after each turn, to force the fluid in the indicator to absorb the gas sample. Then hold the indicator at a 45° angle while you count to five. Level the indicator and immediately note the level of the fluid.

3 Taking the stack temperature. Insert the end of the stack thermometer about three quarters of the way into the flue through the test hole. Clip the thermometer in place and leave it in the hole for three or four minutes, or until the reading has stabilized, before noting the temperature. To determine the net stack temperature, subtract the furnace room temperature from the reading on the stack thermometer.

Using Tests to Set a Burner

Armed with the test readings from your burner (pages 30-31), you can evaluate its performance and make final adjustments to get maximum combustion efficiency. A high carbon-dioxide level and a low smoke density show that the fuel is burning with a high degree of efficiency. To find the percentage of usable heat the burner is producing, use the slide calculator supplied with the carbon-dioxide tester to correlate the carbon-dioxide level with the net stack temperature.

No oil burner is 100 per cent efficient. If the burner is under 10 years old, and equipped with a flame-retention device at the end of the air tube (page 28), or if you have a packaged unit—in which the burner and furnace are designed for each other—you can expect 80 per cent efficiency with a 10 to 12 per cent carbon-dioxide level and a smoke density of no more than 1. If the burner is over 10 years old, or if you have a conversion unit—in which an oil burner has been added to an old system, you usually can achieve between 70 and 75 per cent efficiency with an 8 to 10 per cent carbon-dioxide level and a smoke density of 2 or less.

If your burner does not measure up to these standards, adjust the draft at the draft regulator (page 30) and the burner air supply at the air shutter (below). Retest the draft, carbon-dioxide level, smoke and stack temperature after each manipulation. For safety's sake, pay special attention to the smoke readings. Heavy smoke, which is not always visible, is accompanied by potentially hazardous levels of toxic combustion gases and by heavy soot accumulations that restrict the flue passages—and could eventually even cause a fire. On newer systems, with narrow flues, smoke levels over a reading of 1 may cause the passages to close entirely. Older systems, with larger flues and chimneys, can tolerate heavier smoke, but any reading of 3 or higher probably means that you need to clean the flues at least once a year to get good heat transfer and to prevent blockage.

When you cannot bring your burner up to an acceptable efficiency level simply by adjusting the air supply, check the chart opposite to pinpoint—and correct—other possible causes of trouble. If even these adjustments do not improve your burner's efficiency, your system needs professional attention.

Adjusting the burner air. Loosen the lock nut that holds the air shutter in position. Then open the peephole in the fire door. The flame should burn with a bright, yellowish orange glow and have a slight haze at the flame tips. If the flame is dark and reddish, open the air shutter a fraction of an inch to allow more air into the combustion chamber. If the flame burns with a bright white glow or if hot, gassy-smelling fumes blow out at you, close the shutter enough to cut down some of the air. Do not tamper with the bulk air band unless you have an older burner (right) that is not equipped with a shutter. Adjust the band on an older burner by loosening the lock nut that holds it in place and rotating the band.

After you correct the flame, shut the peephole and make the carbon-dioxide and smoke tests once more. You may have to adjust the shutter and make the tests several times.

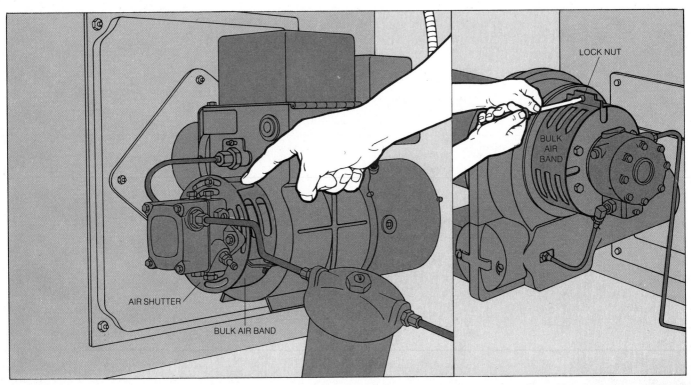

AIR SHUTTER

BULK AIR BAND

LOCK NUT

BULK AIR BAND

Matching Instrument Readings to Oil Burner Problems

Problem	Causes	Remedies
Low overfire draft (below -.01) accompanied by low stack draft (below -.02)	Improperly adjusted draft regulator	Adjust the draft regulator for more draft (*page 30*).
	Leaks in the flue pipe or chimney	Locate and seal leaks (*page 34*).
	Improperly sized or designed chimney	Have a heating contractor check and modify if necessary.
	Blocked flue pipe or chimney	Clean the flue pipe and chimney (*pages 34, 80*).
Low overfire draft (below -.01) accompanied by high stack draft (over -.05)	Soot or other debris restricting the heat exchanger	Check and clean heat exchanger surfaces (*page 25*).
High overfire draft (over -.02)	No draft regulator	Call a serviceman to install a draft regulator.
	Improperly adjusted draft regulator	Adjust the regulator for less draft (*page 30*).
	Insufficient supply of burner air	Adjust the air shutter for more burner air (*page 32*).
	Burner insufficiently ventilated to the outside	Open a window to ventilate the furnace room.
Excess smoke (over No. 1 on burners less than 10 years old, over No. 2 on burners more than 10 years old) accompanied by high carbon-dioxide levels (over 12 per cent)	Insufficient burner air	Adjust the air shutter for more burner air.
	Dirty fan or obstructed air intake openings	Clean the fan (*page 27*).
	Dirty or obstructed air tube	Clean the air tube (*page 28*). If you find damage, call a serviceman.
	Insufficient draft	Adjust the regulator for more draft (*page 30*).
Excess smoke accompanied by low carbon-dioxide levels (less than 10 per cent)	Faulty or damaged nozzle	Replace the nozzle (*pages 28-29*).
	Debris in the combustion chamber	Clean the combustion chamber (*page 34*) and, if necessary, repair the lining (*pages 34-35*).
	Oil spraying onto the end of the air tube or the walls of the combustion chamber	Correct the position of the firing assembly (*page 28*) making sure the nozzle is centered in the air tube.
Low carbon dioxide accompanied by low smoke (carbon dioxide under 10 per cent with 0 smoke on burners less than 10 years old; carbon dioxide under 8 per cent with No. 1 smoke on burners older than 10 years)	Excess burner air	Adjust shutter for less burner air (*page 32*).
	Air leaking into the furnace	Locate and seal leaks (*page 34*).
	Excess draft	Adjust the draft regulator for less draft (*page 30*).
Low carbon dioxide accompanied by excess smoke (above No. 1 on newer burners; above No. 2 on older burners)	Defective nozzle	Replace the nozzle (*page 29*).
	Debris in the combustion chamber	Clean the combustion chamber (*page 34*) and, if necessary, repair the lining (*pages 34-35*).
High net stack temperature (over 600° on packaged units; over 700° on conversion units)	Excess draft	Adjust the draft regulator for less draft (*page 30*).
	Heavily sooted flue pipe or heat exchanger	Clean the flue pipe and the heat exchanger (*page 34*).
	Burner flame too high	Call a serviceman to check the burner, pump pressure and nozzle size.
Low net stack temperature (below 380°)	Burner flame too low	Call a serviceman to check the burner.

Cleaning and Sealing a Furnace

To obtain the best performance from your oil burner, you may have to clean or repair other parts of the furnace. An air leak, a dirty heat exchanger, soot in the flue and the chimney or crumbling and debris-filled combustion chambers all affect burner efficiency.

Most of these problems can be solved quickly and inexpensively. Leaks are sealed easily with furnace cement. The heat exchanger or boiler surfaces of an oil-fired furnace are cleaned in virtually the same way as those of a gas furnace (page 25). You can loosen chimney soot with a bag of rocks as shown on page 80, and then remove it through a cleanout door at the base of the chimney. To dislodge soot from a flue, simply dismantle it and rap each section against a floor covered with newspapers.

A combustion chamber usually gives years of trouble-free service. Eventually, however, intense heat combines with repeated expansion and contraction to crumble firebrick and molded linings; stainless steel simply burns up. If a firebrick lining crumbles in large chunks, you can use furnace cement to bond the pieces back in place. But the best repair for any combustion chamber is to reline it with a preformed, wet liner available from heating and refrigeration suppliers. This type of liner, made of flexible, heat-resistant fibers, is inexpensive and easy to install in any combustion chamber that retains its shape. In many cases, the liner will improve furnace efficiency because it heats up faster than conventional lining materials and reflects more heat.

Before buying a liner, check your furnace or boiler warranty to make sure that you will not void it by modifying your combustion chamber with the addition of a liner. To obtain a liner the right size for your combustion chamber, you will need its dimensions (opposite, top). You will also need the firing rate in gallons per hour, and the spray pattern of your nozzle (page 28). A heating supplier may recommend a different nozzle to compensate for the increased flame intensity caused by the liner.

Stopping Leaks

1 **Locating the leaks.** In the drawing below, the red outlines indicate seams where leaks may occur: at the combustion-chamber cover plate, the burner-mounting flange, the fire door and the flue joints. Leaks may also occur in rusted flue sections. Check for rust by pressing on the flue with your finger. To detect leaks in seams, fire up the burner and move a lighted candle along each seam. The flame will be deflected inward wherever there is a leak.

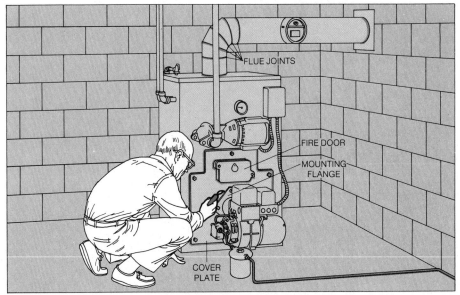

FLUE JOINTS

FIRE DOOR

MOUNTING FLANGE

COVER PLATE

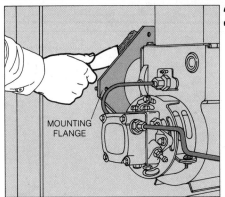

MOUNTING FLANGE

2 **Sealing a leak.** Turn off the burner and allow the furnace to cool. Clean off the surfaces around the leak with a wire brush and use a putty knife to fill leaks with refractory furnace cement. If a flue section is badly rusted and perforated with small holes, replace it. To seal a leak around the burner-mounting flange (left) unscrew the bolts around the edge and pull the flange back a fraction of an inch. Scrape away any old gasket material under the flange and apply a thin layer of cement around the edges. Screw the flange back in place.

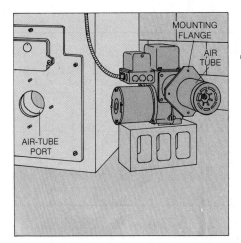

MOUNTING FLANGE

AIR TUBE

AIR-TUBE PORT

Relining Combustion Chambers

1 **Removing the burner.** Shut down the burner at the master switch and service panel, then close the oil-supply valve. Mark the air tube so that you can insert it the same distance into the combustion chamber when you reinstall the burner. Unscrew the bolts on the mounting flange and pull the burner away from the combustion chamber. If you cannot do so without bending the oil-supply line, disconnect the line at the oil-burner pump. As you pull the burner from the air-tube port make sure that any gasket material encircling the air tube does not fall off and break. Set the burner down on its own pedestal or support it on a cinder block.

2 Measuring the combustion chamber. Inspect the chamber by looking in through the air-tube port and by reaching inside and feeling the chamber walls and floor with your hand. Pull debris out through the air-tube port, then vacuum the chamber. To measure the inside length and width of the chamber, hold a tape measure as shown below—with one arm through the fire door and the other holding the tape measure at the air-tube port. If the fire door is too small for you to get your arm in or if it is located so that you cannot reach into the chamber through it, measure the chamber length by inserting a yardstick through the air-tube port. Calculate the chamber width by measuring the width of the furnace and subtracting twice the thickness of the combustion-chamber walls, measured at the air-tube port. Take your measurements to a heating supplier and buy a liner to fit.

3 Preparing the liner. Remove the liner from its plastic bag—it comes wet, so it can be molded—and spread it open. Measure the height of the combustion chamber and the height of the liner. If the liner is taller than the combustion chamber, mark the height of the chamber on the liner and use scissors to cut flaps 4 or 5 inches wide around the top. If the liner does not have a hole for the air tube, measure from the top of the combustion chamber to the top of the air-tube port. Then, the same distance below the top of the liner (or below the flaps if you have had to cut them), cut out a circle slightly smaller than the diameter of the air tube.

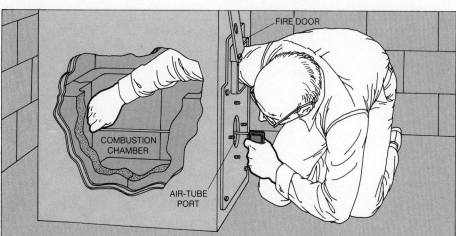

FIRE DOOR

COMBUSTION CHAMBER

AIR-TUBE PORT

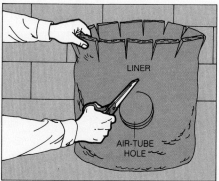

LINER

AIR-TUBE HOLE

4 Lining the chamber. Roll up the liner and push it into the combustion chamber through the air-tube port. Reach through the fire door and the air-tube port and unroll the liner. If you cannot reach through the fire door, you will have to position the liner with one hand. Turn the liner so that the air-tube hole coincides with the air-tube port. Then, working from the back of the chamber to the front, press the liner into place against the walls and the floor of the chamber. Make sure the top of the liner adheres firmly to the top of the combustion chamber. Pat the liner smooth and fold any flaps at the top of the liner over the top edge of the combustion chamber. If you tear the liner, simply press the torn edges back together. You can also patch torn sections with scraps from the air-tube hole.

Partially dry the liner with a propane torch or a light bulb of at least 100 watts until it has the consistency of stale bread. Then, with a sharp knife, trim the air-tube opening in the liner so that its edge is flush with the air-tube port.

Push the air tube into the port up to the mark made in Step 1. Screw the mounting flange to the furnace; reconnect the oil line, if you disconnected it earlier. Open the oil valve and restore power to the burner. If you disconnected the oil line, prime the pump (page 29). Then turn on the burner at the master switch and allow it to run for three minutes and shut it off for three minutes. (You may see a little smoke and detect an unfamiliar odor; both are normal.) Repeat this procedure twice to set the liner.

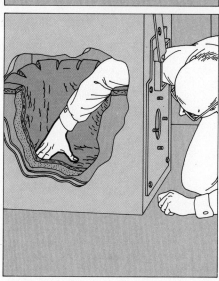

What to Do When Your Burner Won't Start

An oil burner can be balky, stopping unexpectedly or not firing when it should. Before calling a serviceman, try these remedies. In many instances, you can get the burner going yourself.
☐ Check the thermostat setting to make sure it is higher than the temperature in the room.
☐ Make sure you have turned on the master switch—or switches.
☐ Check for a blown fuse or tripped circuit breaker.
☐ Check the storage tank to make sure it is not empty; if it has no gauge, insert a long stick through the filler pipe. Make sure the oil-line valves are open.
☐ Press the restart button on the ignition safety relay; it may be located at the burner or the stack. If you have a stack-mounted relay, open the cover and realign the contacts (page 27) before pressing the button. Do not press the restart button more than once.
☐ Press the restart button on the burner motor.
☐ Remove the firing assembly and check the ignition system (pages 28-29). If the burner still does not start—or starts but shuts off within 60 seconds—call a serviceman.

Preserving the Lungs of Your Forced-Air System

In a forced warm-air system, the blower—or fan—that distributes the heat is spun by a motor that may be attached to the fan shaft *(page 38)* or connected to the shaft by a belt running between pulleys *(below)*. Belts wear out and pulleys get out of alignment, but because they and the motor are outside the fan housing, this system is easy to work on. On the direct-drive system, the motor is inside the fan and less accessible, but repairs are seldom needed.

With either type, the principal problems are noise, too little or too much air flow, and—rarely—a burned-out motor. Vibration noises are often quieted simply by tightening the screws holding the blower housing and the motor, and—with a belt-driven motor—correcting belt tension and pulley alignment. Also check

lubrication of the bearings on the blower and motor. Some bearings are permanently lubricated and sealed, but most require oiling or greasing once a year before the heating season starts—and again at the end of the season if the blower also distributes cool air from a central air conditioner. If air velocity gets so high you hear a siren-like sound from the ducts, cut down the speed *(opposite)*.

If the system fails to deliver enough heat, the problem may be too low a blower speed; speed it up. Sometimes, though, the lack of warm air may be caused by dust and lint. If the blower wheel gets dirty, clean it with a vacuum cleaner and a brush. If the filter gets clogged, replace the glass-fiber type or wash off the plastic or aluminum type.

When a motor burns out, you usually

can replace it with one of the same size. If you have added air conditioning to an old system, however, the new motor probably should be at least a size larger than the burned-out one. To disconnect a motor for replacement, simply undo its wires and remove the bolts holding the motor to its brackets.

Besides keeping the motor and blower running smoothly, you also may want to improve your forced-air system by adding a humidifier *(pages 39-41)* or by installing an electrostatic filter. Whatever the addition or adjustment you plan to make, always shut down the power completely before you tackle the job. Turn off the furnace master switch and the circuit breaker or remove the fuse bringing power to the furnace. Be sure to leave the power off until you finish working.

Belt-drive Blowers

Lubricating the motor and blower. Unsnap or unscrew the access panel in front of the blower unit. If the bearings have no oil cups or grease fittings, lubrication is not needed. If you see cups, lift the lids or pull the plugs one at a time at both ends of the motor and blower. Dribble six to eight drops of 10- to 20-weight nondetergent electric-motor oil, available at hardware stores, into each cup. If you see grease fittings—the hexagonal nipples ending with flattened balls that are common on blower bearings—use a hand-operated grease gun to lubricate them with two full pumps of automotive lubrication grease.

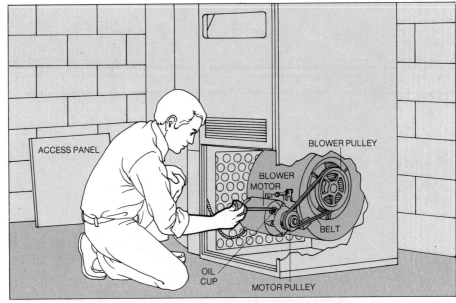

Aligning the pulleys. Set one edge of a carpenter's square against the outside faces of the motor and blower pulleys to make sure they are positioned in a straight line and at right angles to the motor shaft. If the pulleys are less than ½ inch out of alignment, use a hex wrench to loosen the setscrew holding the motor pulley to the motor shaft, and slide the pulley back or forward as necessary. Retighten the setscrew. If the setscrew is rusted or if you need to move the pulley more than ½ inch, loosen the bolts holding the motor to the mounting bracket. Slide the motor along the bracket until the pulleys align, then retighten the bolts.

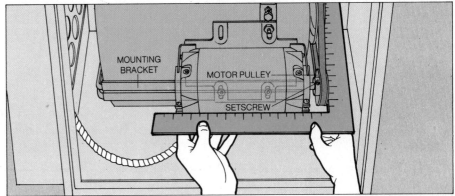

Checking the belt. With your hand, press on the belt midway between the motor and blower pulleys. If the belt deflects more than about ¾ inch up or down, use a wrench to turn the motor adjustment bolt clockwise to increase tension. If the belt deflects less than ¾ inch, turn the bolt counterclockwise to decrease tension.

Replace a frayed, stringy or cracked belt immediately—if it breaks it may cause damage. Any A-type V belt of the correct length will do, whether from a heating-supply dealer, an auto-supply store or a garage. Turn the motor adjustment bolt counterclockwise until the belt slips easily off the pulleys. To attach a new belt, loop one end around the motor pulley. Slide the edge of the other end of the belt onto the blower pulley. Rotate the blower pulley by hand until the belt feeds into the slot. Correct the tension when you put the belt on and check it again in two weeks; new belts often stretch.

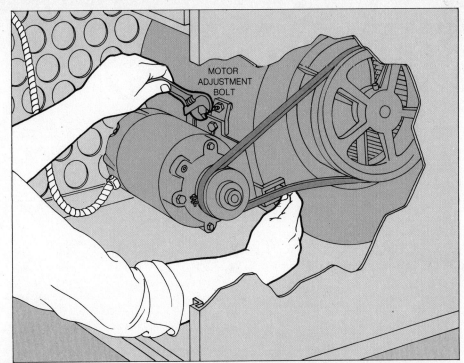

MOTOR ADJUSTMENT BOLT

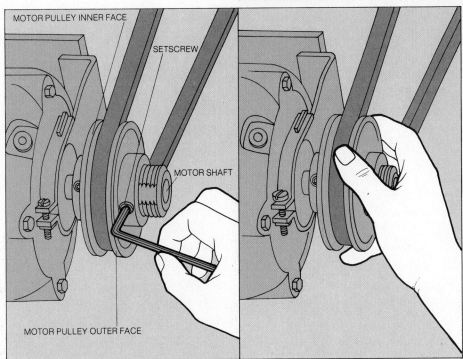

MOTOR PULLEY INNER FACE

SETSCREW

MOTOR SHAFT

MOTOR PULLEY OUTER FACE

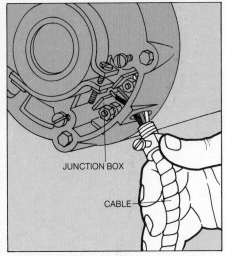

JUNCTION BOX

CABLE

Removing the motor. Slide off or unscrew the plate covering the junction box at the end or side of the motor. Detach the wires from the electrical terminals inside the box, taking note of which terminals they attach to; unscrew the lock nut holding the armored cable that covers the wires and pull the cable and wires away from the motor. Loosen the lock nut on the motor adjustment bolt, turn the bolt counterclockwise to release the belt tension and slip the belt off. Remove the bolts holding the motor to the mounting bracket and lift off the motor.

To replace the motor with the same or a larger model, fasten the new motor to the old bracket and slip the belt over the pulleys. Then attach the cable to the motor junction box and secure the wires to the terminals in the original arrangement. Cover the box and adjust the belt.

Changing blower speed. With a hex wrench, loosen the setscrew locking the two faces of the motor pulley together (above, left). To increase blower speed, turn the outer pulley face clockwise in increments of 180° to bring it closer to the inner face (above, right). In some cases you may need to release the belt tension to rotate the face of the pulley. If adjusting the pulley until the faces touch does not provide enough speed, replace the motor pulley with another adjustable one of the next larger diameter. To decrease blower speed, turn the outer pulley face 180° counterclockwise to separate it from the inner face. Retighten the setscrew against one of the two flat spots on the outer end of the shaft; readjust the pulley alignment and belt tension.

Caution: Increasing the speed raises the amperage, or amount of current, going into the motor and may burn it out. About an hour after you adjust the pulleys, check the motor. If it feels unusually hot, decrease the speed.

Direct-drive Blowers

Oiling the motor and blower. Remove the access panels from both the blower and furnace compartments. Use a screwdriver to loosen the two front screws in the metal strip between the compartment openings. Slide the blower partway forward by pulling the metal shelf below the unscrewed strip. If the wires are too taut to permit the blower to slide easily, unclip the wires from along the side of the furnace opening or detach or unplug them at the blower or furnace junction box. Look for oil cups *(inset)* at the visible end of the blower motor; if there are none, lubrication is not needed. If you find oil cups, lift the lids or pull the plugs and drip six to eight drops of 10- to 20-weight nondetergent electric-motor oil into each cup. Slide the blower back in place, reattach the metal strip and put back the access panels.

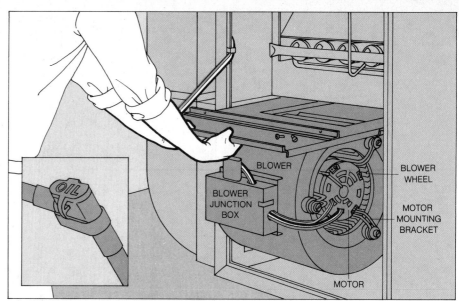

Adjusting a multispeed motor. Whether the motor has a blower-mounted junction box *(right)*, an exposed terminal strip, or a junction box beside the furnace box, only one or two of the terminals on the board inside the box or on the strip will be wired with hot lines from the main power source. One hot line (usually red) will be connected to the low or medium-low terminal that powers the blower when the furnace is in operation; the other hot line (usually black) will control the high and medium-high terminals only used for central air conditioning. The neutral line (usually white) will be wired to a neutral terminal. To increase or decrease blower speed, unplug the appropriate hot line and attach it to the adjacent terminal, then replace the junction-box cover and access panel.

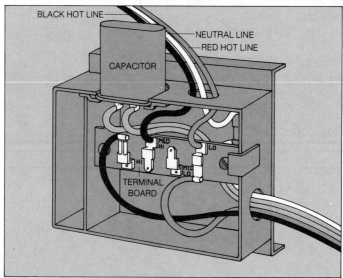

Replacing the motor. Disconnect the wires to the blower by unplugging the box or the hot and neutral lines. Then slide the blower and attached shelf *(right)* onto the floor. Remove the bolt assemblies holding the motor mounting bracket to the blower housing and loosen the bolt that connects the opposite end of the motor shaft to the blower wheel. Ease the motor out of the blower. Remove the nut and washer at the ends of the bracelet-like ring of the mounting bracket *(far right)*. Slip off the bracket and save it.

Fit the bracelet around the new motor approximately as far from the ends as it was on the old motor. Slide the motor into the blower, reattach the mounting bracket and tighten the bolt against the flat spot on the motor shaft. Rotate the blower wheel by hand; if the wheel rubs the housing, loosen the bolt and shift the wheel sideways until it rotates freely. Slide the blower back in place and reconnect the wires. Reattach the access panel.

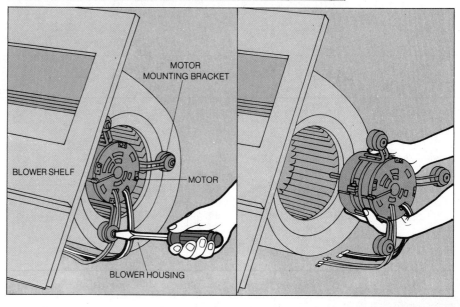

Relief from Dry and Dusty Air

In winter most homes get too dry for comfort. But with forced-air heat you can keep every part of your home at the moisture level that is generally considered pleasing—a relative humidity of 30 per cent—by adding a central humidifier to the system. And if dust and pollen control are important considerations, you can also provide cleaner air by replacing the existing filter in the forced-air system with an electrostatic air filter.

The most popular type of central humidifier passes the heated air from the furnace through a wet pad or disk before the air is distributed through the house. Most wetted-element humidifiers are designed to fit, as shown here, on the warm-air plenum connecting the furnace with the duct network. Some models, however, call for a bypass pipe between the plenum and the cold-air return duct, or they attach to a duct leading out of the plenum. All are installed similarly: cut a hole in the plenum or duct, attach a support bracket (called a stiffener), mount the humidifier on the stiffener and connect it to a cold-water pipe. Mount the humidistat control—which usually comes with the unit—on the return air duct, and wire the control and the humidifier to an electric power source. Take the power from the furnace junction box if the units use 120 volts or from the furnace transformer if the units require 24 volts.

Electrostatic filters are installed in much the same way: cut a hole in the return air duct, set in the unit and then wire it to the furnace junction box. Detailed directions for installing a humidifier are shown on these pages; consult the manufacturer's instructions for installing an electrostatic filter. Most electrostatic filters come in one standardized size, but humidifiers are sold in a wide range of models. Humidifier capacity is generally expressed in the number of gallons of water a day the humidifier will use. Where the capacity is given in pounds of water a day, you can convert the figure into gallons by dividing it by 8.34. The humidifier size you need (chart, right) depends on where you set your thermostat, the size of your house and how well the house is weatherproofed.

Estimating Your Humidity Needs

Thermostat setting	Weatherproofing of house	Gallons a day for each square foot
65°	Tight	.003
	Average	.004
	Loose	.005
70°	Tight	.003
	Average	.005
	Loose	.006
75°	Tight	.004
	Average	.007
	Loose	.009

Sizing a humidifier. To determine the humidifier capacity you need, first check the weatherproofing of your house against the definitions of tight, average and loose houses on page 124. Then read across the chart above from the thermostat setting you usually use and multiply the figure in the right-hand column by the total number of square feet on all the floors of your house—including the basement. This chart is based on an average ceiling height of 8 feet. The figures that are shown here also are computed for a wide variety of climatic conditions, which accounts for the apparent discrepancies in some instances.

Mounting a Humidifier Bracket

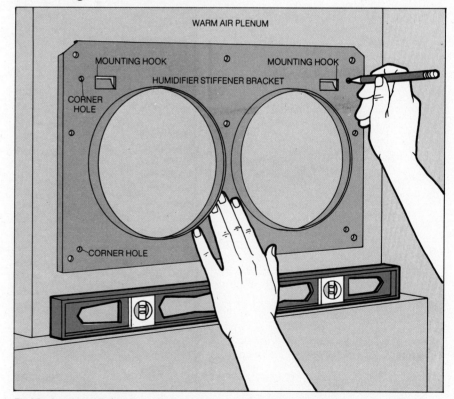

Positioning the stiffener. Shut off all power to the furnace. Locate the stiffener, or humidifier support bracket, on the warm-air plenum; if your forced-air system includes an air conditioner, place the stiffener on the side of the plenum parallel to the length of the A-frame coil and above the condensate pan. Level the stiffener, then mark onto the plenum the locations of all screw holes and of the corner holes for the cutout in the plenum. Connect the corner marks. To attach the stiffener, use a drill or an awl to make screw holes in the plenum. Then make slits large enough for a saber-saw blade or metal shears by drilling a series of holes at each corner for the rectangular cutout or by punching through the plenum with a screwdriver (page 51). Saw or cut out the rectangle along the drawn lines. Screw the stiffener to the plenum.

Hooking Up the Humidifier

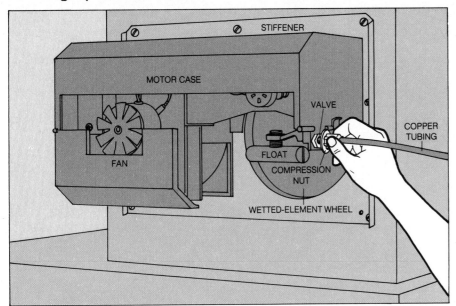

1 **Mounting the humidifier.** Separate the motor case of the humidifier from the water reservoir below it. Set the reservoir aside. Fit the case onto the hooks on the stiffener bracket and insert the metal screw that secures the bottom of the case to the center of the stiffener. To prepare the humidifier for the water connection, slip the nylon compression nut and sleeve provided with the unit over one end of the copper tubing. Then push the end of the tube into the humidifier valve at the side of the case and tighten the compression nut with your fingers.

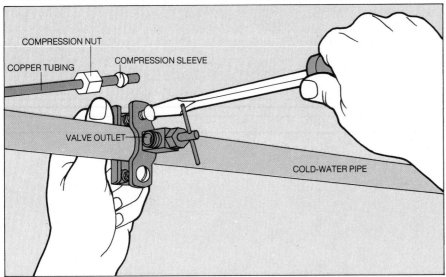

2 **Hooking up the water supply.** Cut off the house water supply by closing the main shutoff valve, and drain the system *(page 42)*. If the cold-water pipe nearest the humidifier is copper or plastic, you can use a self-tapping saddle valve *(left)* to tap into it. Remove one of the bolts on the saddle-valve clamp, place the rubber pilot washer on the stem of the valve, and slide the clamp onto the pipe. Then replace the bolt in the clamp, tighten it securely and turn the valve handle clockwise until the stem pierces the pipe. If the nearest cold-water pipe is steel, use a plain saddle valve and bore a hole for the valve stem with a hand drill or a properly grounded electric drill. Uncoil the copper tubing and slip the brass compression nut and sleeve over the loose end. Insert the end of the tubing into the saddle-valve outlet and tighten the nut first with your fingers, then with a wrench.

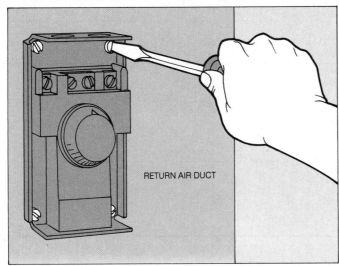

3 **Mounting the humidistat.** To insert the temperature sensor of the humidistat in the return air duct, position the humidistat template at a convenient height and as near as possible to a 120-volt junction box or 24-volt transformer, depending on the kind of power your humidistat requires. For a horizontal or downflow furnace, locate the template 6 to 8 feet beyond the blower inlet on the duct. If you do not have a template, draw a rectangle slightly larger than the humidistat sensor onto the duct. Drill the corners of the rectangle and saw or cut out the sensor hole, following the same method you used to cut the hole in the plenum *(page 39)*. Carefully slide the sensor into the hole and push the humidistat case flat against the duct. Remove the cover of the case, drill holes for the mounting screws and secure the humidistat to the duct.

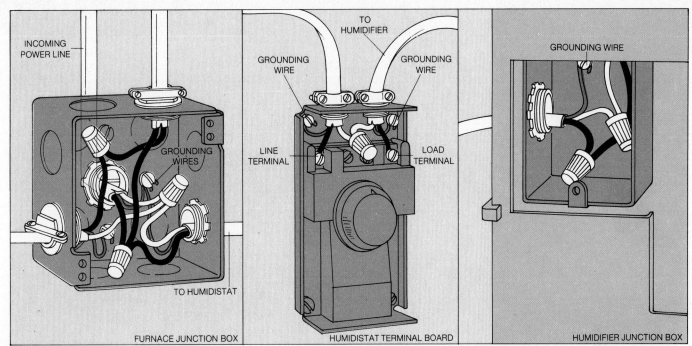

INCOMING
POWER LINE

GROUNDING
WIRES

TO HUMIDISTAT

FURNACE JUNCTION BOX

TO
HUMIDIFIER

GROUNDING
WIRE

GROUNDING
WIRE

LINE
TERMINAL

LOAD
TERMINAL

HUMIDISTAT TERMINAL BOARD

GROUNDING WIRE

HUMIDIFIER JUNCTION BOX

4 **Hooking up the electricity.** For an all 120-volt system, connect the humidistat and humidifier to a 120-volt junction box—usually the one on the furnace—with two lengths of No. 16 two-wire grounded plastic-sheathed cable. Attach one end of the first cable to the box, matching the white wire to the white or neutral wire of the incoming power line, the black wire to the black or hot wire on the same line, and wrapping the ground wire around a mounting screw (*above, left*). Feed the other end of the first cable through the access hole of the box into the hole in the humidistat, wrap the ground wire on a mounting

screw and attach the hot line to the line terminal—leaving the neutral line dangling. Then feed one end of the second cable into the humidistat, wrap the ground wire on a mounting screw and attach the hot line to the load terminal. Twist the ends of the two dangling neutral lines together and screw a wire cap over them (*center*). Remove the humidifier junction-box cover and slip a Romex connector over the loose end of the second cable. Feed the end of the cable into the box through the access hole and secure it with a connector washer. Ground the paper-sheathed wire, then attach the hot and

neutral lines to the terminal board or loose wires in the box (*right*). For an all 24-volt system, use the same wiring plan but get the power from the 24-volt transformer usually mounted at the furnace junction box and use No. 18 thermostat wire. If there is no 24-volt transformer, install one at a 120-volt junction box, following the techniques on pages 17 and 18, Steps 1-3. For a dual voltage system, use the same wiring plan to connect the humidistat and humidifier to a 24-volt transformer. Plug the prewired 120-volt line into a grounded outlet or run a 120-volt line from the 120-volt box to the humidifier.

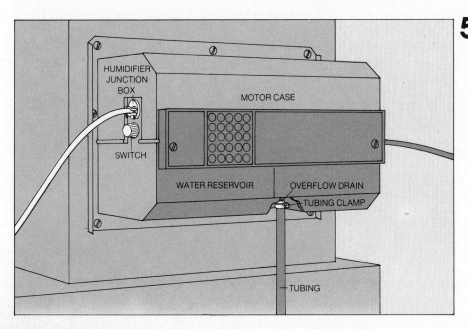

HUMIDIFIER
JUNCTION
BOX

MOTOR CASE

SWITCH

WATER RESERVOIR

OVERFLOW DRAIN

TUBING CLAMP

TUBING

5 **Attaching the overflow drain.** Slide the water reservoir of the humidifier up under the motor case and attach it to the case. Fit one end of the drain tubing provided with the unit over the projecting sleeve of the overflow drain on the reservoir; secure the tubing with a tubing clamp. Run the other end of the tubing to a nearby sink or basement floor drain.

To start the humidifier, turn on the house water supply at the main valve. Then flip on the furnace master switch and restore power to the line at the circuit breaker or fuse box. Set the humidistat to the relative humidity level desired and turn on the humidifier switch.

Hot-Water Heat: Easy to Maintain and Modernize

Keeping a forced hot-water heating system running smoothly takes only a minimum of regular servicing. And bringing the system up-to-date with new valves *(page 44)* or new convectors *(pages 45-47)* is simple piping work.

Once a year, before the boiler starts up, bleed the radiators or convectors *(page 9)*. At the same time lubricate the pump that circulates the water through the system by putting a few drops of No. 20 nondetergent electric-motor oil into the oil cups at both ends of the motor and on the top of the bearing assembly between the motor and pump body.

During the heating season make periodic checks of the water pressure in the system by examining the combination gauge mounted on the side or front of the boiler. Depending on the size of your house, pressure can safely range from as little as 3 pounds per square inch, when the water cools and contracts, to about 30 pounds when it heats and expands.

If the movable "pressure" pointer on the gauge drops below the stationary "altitude" pointer, you need to increase the pressure in the system. How you do this will depend on the type of expansion tank you have *(page 44)*. The tank, which provides a cushion of air for the expanding and contracting water, may be a conventional one, with a top layer of air in direct contact with a layer of water, or it may be a diaphragm tank, with the air layer at the bottom, separated from the water by a rubber membrane. With the conventional expansion tank, you increase the pressure by adding water to the system; with the diaphragm tank, you recharge the air in the tank instead.

If there is too much pressure—i.e., when the movable pointer nears the 30-pound mark—it should be lowered. You can recharge a conventional tank with air *(page 44)*, but call a serviceman if you have a diaphragm tank.

With care you can keep the repair jobs few. Components rarely fail, but when breakdown occurs it most likely involves one of these units: the pump motor, the coupler holding the motor to the pump shaft or the pump seal. Remove the burned-out or broken part *(opposite)* and take it to the heating-supply dealer so you can be sure to get an identical replacement. If you must replace the pump seal, or change the valves or heating elements, first drain the system *(below)* and then refill it.

Draining and Refilling the System

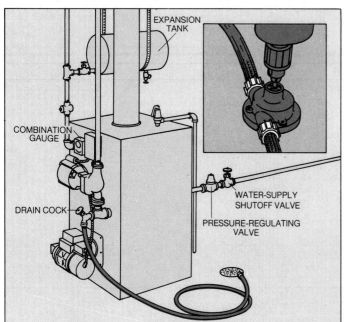

EXPANSION TANK

COMBINATION GAUGE

DRAIN COCK

WATER-SUPPLY SHUTOFF VALVE

PRESSURE-REGULATING VALVE

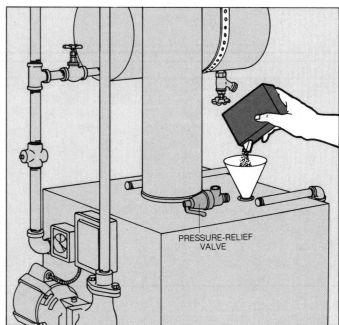

PRESSURE-RELIEF VALVE

1 **Draining the water.** Cut off the boiler at the master switch and service panel. When the water in the system cools to lukewarm as shown on the combination gauge, close the water-supply shutoff valve. Attach one end of a garden hose to the boiler drain cock and run the other end of the hose to a floor drain (or use an electric-drill pump such as the one shown in the inset to lift the water to a sink or an open window). Open not only the boiler drain cock but also the bleeder valves of all the radiators or convectors on the upper floor of the house.

2 **Adding water.** Close the drain cock. Unscrew the pressure-relief valve and pour rust inhibitor into the opening. Replace the relief valve. Close the radiator or convector bleeder valves. Then open the water-supply valve. If there is a pressure-regulating valve on the line, the flow will stop automatically when the system is full. Otherwise, fill until the movable pointer corresponds to the positon of the stationary pointer. Bleed all of the heating units on the upper floor. If you do not have a pressure regulator, have someone bleed each unit while you watch the gauge.

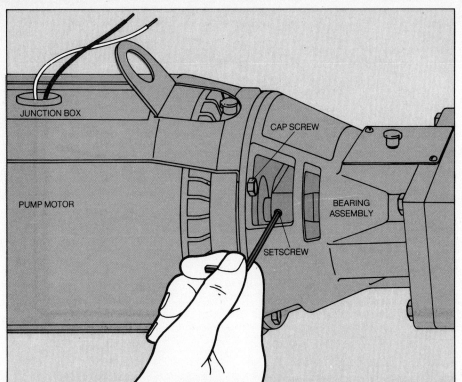

Replacing Pump Parts

Replacing the motor. After turning off the power, remove the junction-box cover from the pump motor and disconnect the wires. Use a hex wrench to remove the setscrew holding the coupler to the motor shaft. Then, gripping the motor in one hand, use an open-end wrench to loosen the cap screws that hold the motor to the bearing assembly. Back the motor out, leaving the coupler attached to the pump shaft.

To install the new motor, fit the free end of the coupler onto the motor shaft. Hold the motor against the bearing assembly, reinsert the cap screws and secure the coupler setscrew.

Replacing the coupler. Remove the motor and use the hex wrench to loosen the setscrew holding the coupler to the pump shaft. Slide the coupler off. To install the new coupler, secure one end to the pump shaft with the setscrew. Then replace the motor and attach the other end of the coupler to the motor shaft.

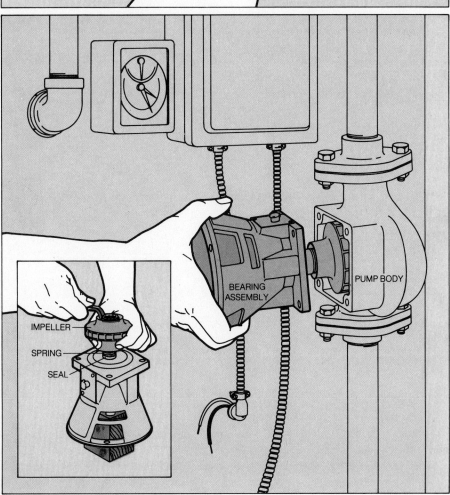

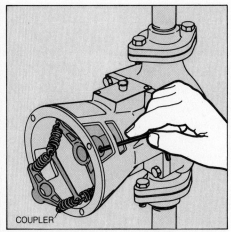

Replacing the pump seal. Drain the system (*opposite, Step 1*) or turn off the power to the burner and then cut off the water to the pump by shutting the valves above and below it. Remove the motor and coupler, then undo the cap screws holding the bearing assembly to the pump body. Pull the assembly out. Stand the bearing assembly on a wood block to support the pump shaft (*inset*). Turn a box or socket wrench clockwise to loosen the nut holding the impeller to the pump shaft. Slide off the impeller and spring, and save them. Pull off the brass seal.

To install the new seal, slide it onto the shaft and press it tight. Attach the old spring and impeller with the nut and washer. Reassemble the pump. Refill the system (*opposite, Step 2*) or open the valves at the pump. Restore the power.

Recharging an Expansion Tank

Recharging the expansion tank. If your system has a conventional tank, run a garden hose from the combination valve at its base to a floor drain or sink. Then close the shutoff valve on the line between the tank and the boiler and open the combination valve until the tank is emptied. If there is no combination valve, open the drain and let the water empty into buckets; if you do not have a shutoff valve, drain and refill the entire system (*page 42*). If your system has a diaphragm tank (*inset*) check the pressure by attaching a tire gauge to the air-recharge valve. If air is needed, use a bicycle pump to add it.

If you want to install a combination valve do so while the tank is empty. Close the shutoff valve. Then use a hacksaw or tubing cutter to cut the vent tube of the valve to two thirds the height of the expansion tank. Using a pair of open-end wrenches, remove the plug or drain cock from the base of the tank and screw the combination valve into the opening. Open the shutoff valve.

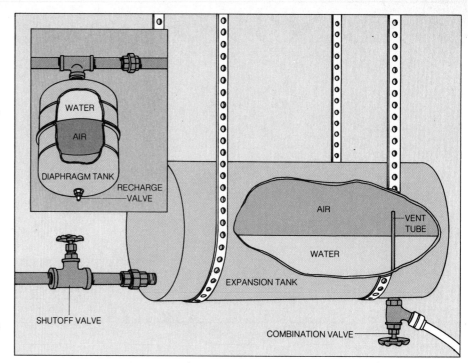

Adding a Shutoff Valve

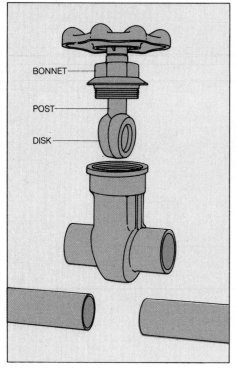

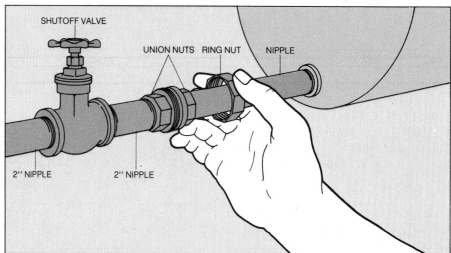

A sweated fitting of copper. Drain the system (*page 42, Step 1*). Working on the line between the expansion tank and the boiler, use a tubing cutter or hacksaw to cut out a section 1 inch shorter than the length of the shutoff valve. Before soldering it in, use a wrench to unscrew the bonnet from the valve body and lift out the disk assembly; otherwise the soldering heat may warp the disk or post.

A threaded fitting of steel. Drain the system (*page 42, Step 1*). Hacksaw through the line near the inlet to the expansion tank. Unscrew the pieces of cut pipe from their fittings at both ends of the section. Attach a 2-inch nipple—a short pipe with threads at both ends—into one existing fitting. Then attach a shutoff valve and another 2-inch nipple. Screw on an assembled union and measure from the union to the remaining existing fitting to determine the length for the third nipple. Undo the ring nut of the union and lift off the free union nut. Then attach the third nipple to the existing fitting, slip the ring nut over the nipple and attach the remaining union nut. Slide the ring nut over the union nuts and tighten it. Refill the system (*page 42, Step 2*).

44

New Convectors for Old Radiators

The old-fashioned radiator, which warms partly by radiation (by giving off heat rays) and partly by convection (by warming cold air that flows up over it), is being largely replaced nowadays by the less obtrusive convector, which, as the name indicates, works mainly by convection. Both kinds of units have a pipe at each end, connecting to the supply and return main from the boiler. One pipe, called a supply riser, brings water into the unit; the second pipe, called a return riser, carries water back to the main. Because the piping systems for both are identical, you can take an old radiator out and use its risers for a new convector.

Convectors come in either upright or baseboard models, in either copper or steel, and with either aluminum or cast-iron fins. Copper with aluminum fins is easiest to install and comes in a wide range of sizes to suit your heating needs (pages 124-125). Because metals heat and cool differently, however, mixing even one or two copper units with existing iron ones may make some areas too hot while others are too cold.

The piping required for substituting convectors is minimal. Some upright units may align so exactly with old risers that you can connect them with unions and nipples (right). For most uprights and all baseboard convectors you can zigzag new stretches of pipe to the joints downstairs where the risers elbow up from the main (overleaf). Where codes allow, copper is the easiest piping material to use—Type L or M rigid tubing for straight stretches and flexible connectors for turning corners. Where the new copper meets the old steel piping or the threaded ends of the pipes on upright convectors, install steel-to-copper adapters or transition fittings.

Upright convector pipes come fitted with tappings for bleeder valves. To vent a baseboard unit, use an elbow with a vent tapping to connect the baseboard pipe to the return riser. Similarly, use an angle valve at the supply riser so you can balance the heat of the baseboard (pages 12-13). For the upright convector, you can install a shutoff valve (page 44) on the supply riser line.

Removing a radiator. Drain the system (page 42, Step 1) and open the radiator's air vent. Pull back any carpeting, then cover the floor with heavy tarpaulin or plastic. Set a pan nearby to catch the residual water—usually ink black—that will drip from the radiator when you disconnect it. Using a pipe wrench, loosen the union between the radiator and the elbow at each side of the unit. To break tight joints, fit a length of pipe over the wrench handle for extra leverage. Plug the openings of the radiator with rags, tip it onto an old rug turned pile-side down, and drag it out. Then disconnect the elbows from the supply and return risers.

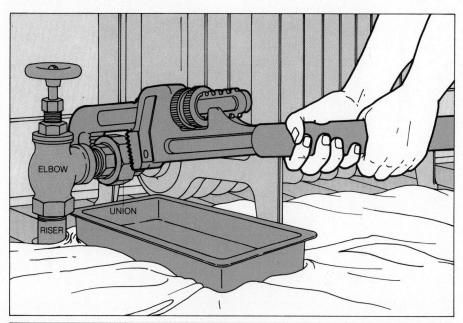

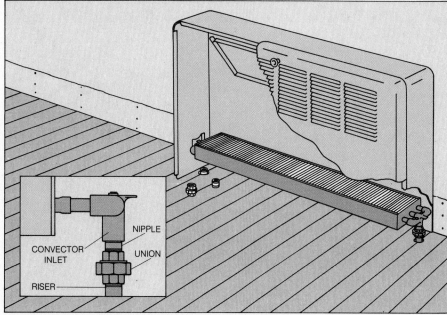

Connecting an upright convector. If the inlet and outlet of the convector align with the existing radiator risers, set the unit in place and center the inlet and outlet above the risers. Screw an assembled union onto each riser tip—using a reducing bushing in the union if the convector is smaller than the riser pipe. Measure from the top of each union or bushing to the convector inlet or outlet above it and add 1 inch to determine the length for each nipple you will need to connect the convector to the risers. Attach the nipples and half unions to the convector and the other halves of the unions to the risers (inset), using the technique for making a threaded fitting shown on the opposite page. Remove the plug from the tapping over the return riser and screw in a bleeder valve. Then trace the supply riser down to the basement or crawl space and add a shutoff valve at a convenient place on its line. Refill the system (page 42, Step 2).

An Upright Convector

1 **Preparing the risers.** Trace the supply and return risers from the convector location down to the basement or crawl space where they meet the supply main. Use pipe wrenches to remove the unions holding the risers to the Ts on the main. Pull the risers out and take each of them apart by disconnecting the elbow farthest from the union. Reconnect the remaining sections of each riser to the union. Then attach transition fittings to the free ends of both risers. If the riser pipes are larger than the convector pipe, use a reducing transition fitting.

2 **Making new riser holes.** Set the convector in place and attach a transition fitting to each end of the heating element. Use a string to mark the floor under the center of each fitting, then draw a circle 1 inch larger than the convector pipe at each mark. Drill pilot holes through the floor and poke a wire through them to make sure the circles are not over joists. Use an electric drill with a hole cutter to cut out both circles.

3 **Connecting the convector.** At each riser, measure from the convector through the hole to the transition fitting on the horizontal riser section. Feed one end of a flexible copper connector of approximately this length down through each riser hole and sweat the top of the connector to the transition fitting at the convector. To make a connection through a joist as shown, draw a vertical line on the joist close to the connector. Then drill a hole 1 inch larger than the diameter of the connector at the midpoint on the line. Feed the connector through the hole. Finally attach the free end of the connector to the transition fitting on the riser and sweat the joint. Remove the plug from the tapping in the convector located over the return riser and screw in a bleeder valve. Add a shutoff valve (*page 44*) on a steel-pipe section of the supply riser. Then refill the system (*page 42, Step 2*).

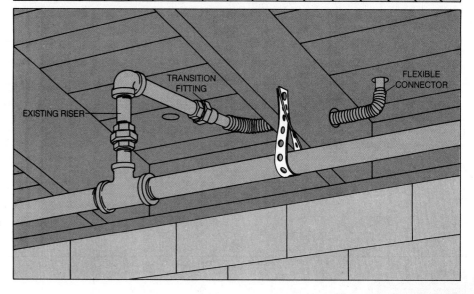

A Baseboard Convector

1 Preparing the wall. Remove the protruding sections of the old risers and attach transition fittings to each horizontal riser section (*Step 1, opposite*). With a pry bar, pull off first the base-shoe molding and then the baseboard from the entire wall. On the floor, mark the location of each stud. If you cannot see the studs at the base of the wall, run a file along the floor to locate them, or drill pilot holes into the wall and use a wire to find them.

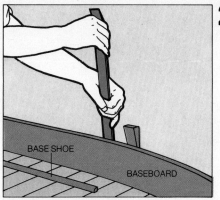

BASE SHOE

BASEBOARD

2 Positioning the convector. Set the back panel of the baseboard convector in place. Slide heating-element brackets onto the back panel—placing them at the studs nearest the ends of the convector and at every second stud in between. Seat the element on the brackets. Sweat a 90° elbow equipped with an air-vent tapping onto the return end of the heating-element pipe. Then remove the disk assembly from an angle valve and sweat the valve onto the supply end of the pipe. When the pipes cool, screw in a bleeder valve and replace the disk assembly of the angle valve. Locate and cut holes for the new riser sections (*Step 2, opposite*).

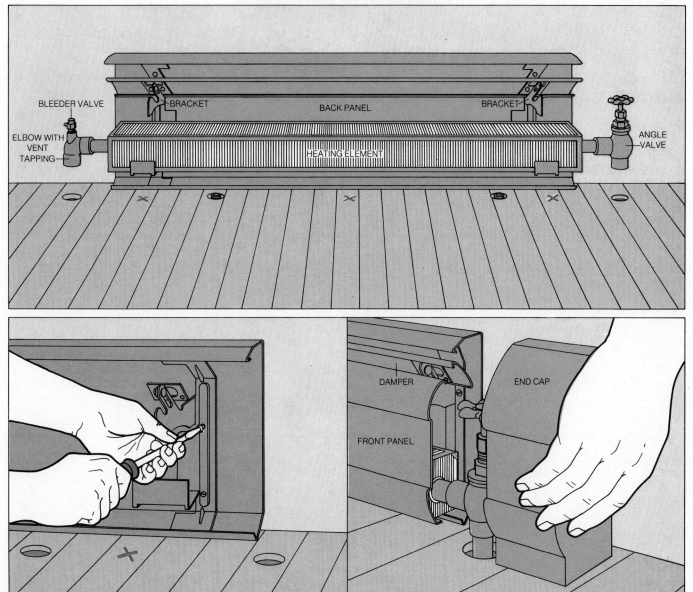

BLEEDER VALVE

ELBOW WITH VENT TAPPING

BRACKET

BACK PANEL

BRACKET

ANGLE VALVE

HEATING ELEMENT

DAMPER

FRONT PANEL

END CAP

3 Mounting the convector. Remove the heating element from the brackets. To attach the back panel to the wall, drill through the top and bottom of each bracket (at the prepunched holes if the bracket has them) and insert 1-inch screws into the studs. Replace the element and use the techniques in Step 3, opposite, to connect the elbows at the ends of the element to the transition fittings on the old riser sections. If the new convector is much wider than the old radiator, you may need straight pipes and elbows to reach from the connectors to the old risers. Attach the damper and front panel to the baseboard and cover the ends with caps or wall-trim fittings. Then refill the system (*page 42, Step 2*).

The ancient heater updated. This freestanding fireplace, one of many shapes and sizes available in almost any color imaginable, comes as a set of ready-made parts that quickly fit together. Firebox, hearth and damper are mounted on a fireproof base, then the various chimney sections are joined and run up to the roof either through the house or along the outside.

People are always adding onto their houses—enclosing a porch, finishing a basement or an attic—and they expect the new living spaces to be kept, like the rest of the house, at shirt-sleeve comfort over a North Dakota winter. An easy way to meet that expectation—or to add heat to a chilly room—is to extend an existing central system, or in some cases, to install one or more independent space heaters, which are available in astonishing variety.

Extending the system you already have offers a number of advantages. For one thing, you need not buy any new heat-producing equipment; you simply get more use out of what you already have, generally with little or no increase in fuel consumption. Moreover, you can upgrade as well as extend an existing central system. If you have forced-air heat, for example, tapping into existing ducts not only lets you channel heat where you need it, but may also permit you to add extra registers for better circulation. In the case of a combined heating-cooling system you can match seasonal needs by installing low registers for heating and high ones for cooling.

Forced-air systems are the simplest to extend. The thin metal used is nearly as easy to bend as cardboard (pages 50-53), and no watertight joints are required. Ducts pose some problems, however. You cannot lead one around a corner without losing much of its heat-delivering capacity. And ducts are bulky; you may have to conceal them in closets or framed enclosures. The pipes of a hot-water or steam system, while requiring threaded or soldered connections, are seldom more than 1 inch in diameter, are easily set inside walls or under floors and adapt gracefully to turns (pages 66-69).

If your central system is already working to capacity or cannot readily be extended to the space you want to heat, the answer may be an independent heater. Electric space heaters are the easiest to install. Oil and gas heaters cost less to run but are usually more obtrusive and may be more trouble to put in. Even fireplaces, when properly used (they seldom are), can be effective (pages 78-81). There are also ingenious and economical ways to steal or borrow heat. A flue heat exchanger captures some of the warmth normally wasted up a chimney (pages 60-61). With a device that combines a pump, a fan and a convector you can warm a room by diverting a little of the hot water generated for other purposes (pages 70-71).

In most of the United States solar heating cannot fully replace other forms of heat energy, but it can at least reduce heating costs. You can install a solar heater for a swimming pool or for domestic hot water yourself (pages 86-91), but choose equipment carefully from among the many kinds being offered. Try to get a model that you know has operated successfully for several seasons in your area.

Extending the Flow of a Forced-Air System

If your house has forced warm-air heating, minor surgery on the existing ductwork can bring heat to a new living space or make an old one warmer. In some cases you need only add a register—a new opening into the room. In others, you may have to extend existing ducts or run new ones. Most of the work is elementary carpentry, and the metal of ducts and connections is cut and shaped like very stiff cardboard (opposite).

Both registers and ducts must be installed according to a plan. A furnace and its ducts resemble a tree trunk and its branches. Some provide the supply, carrying warm air to the rest of the house. Others are returns, taking cool air back to the furnace to be heated. Before you modify an existing branch or add a new branch to the system, find out where all the ducts run and what they do.

Start at the furnace. While it is running, touch the ducts connected directly to it; supply ducts will feel warm, returns cold. Other exposed ducts in basement or attic are identified the same way.

Now go to the registers. Identify the ones that blow warm air in and the ones that suck cool air out. Remove the registers and examine the spaces behind them. Each register connects either directly to a duct or to a fitting called a boot that makes the duct connection; usually, you can see whether the duct runs down to the register, up to it, or horizontally. For a final check on the routes of supply ducts, cool your house to 50° on a cold day, then set the thermostat at 70°. After about 10 minutes feel the ceilings, walls and floors, tracing the warm surfaces that reveal hidden ducts.

With the map of your system completed, you can choose the points at which to tap it with new registers or extend it with new ducts. Adding new registers to existing supply or return ducts (pages 52-53) can make a room more comfortable in all weathers. A return register added to a room that does not already have one can improve air circulation in a heating system. (Returns are not installed in bathrooms and kitchens to avoid spreading moisture through the house.)

In a house with a combined heating-cooling system, a high or low supply register added to complement an existing one enables you to match the circulation to the season. Close the low register in summer and cool air flows down into the room; close the high register in winter and warm air rises from the low one.

Adding new ducts (pages 54-59)—usually to heat a finished attic, a basement room or an addition—is simple if you run the new ducts through a room but outside a wall or ceiling rather than running them inside. As you map your existing system, locate boxed-in places—a closet, for example, or a suspended ceiling—and use these spaces for ducts to save the work of building enclosures.

Such alterations to an existing system have their limitations, particularly in extensions to new living spaces. If your house now seems chilly in winter and your furnace runs almost constantly, it may not be able to heat additional space efficiently. But most heating plants have reserve capacity that is never called upon—in the average forced warm-air system, the reserve is enough to heat two medium-sized bedrooms, or a basement recreation room plus a small attic bedroom, or a garage converted to living space—and all you need do to take advantage of this untapped heat is provide ducts and registers. If you are uncertain that sufficient reserve is available, it is best to consult a heating contractor. For a modest fee he will analyze your system to make sure that your furnace and blower can handle an added load—and advise you on the best route for new ducts.

Working with Sheet Metal

Cutting a round duct with a hacksaw. An open round duct, or pipe (page 56), can easily be shortened with a pair of tin snips, but the closed pipe shown at right must be sawed. Set the pipe, seam upward, in a snug cradle of 2-by-4s nailed on edge to a piece of plywood. Saw through the seam, rotate the pipe slightly and saw again from the top down. To keep the saw from binding, continue to rotate the pipe each time the saw cut penetrates about 1 inch.

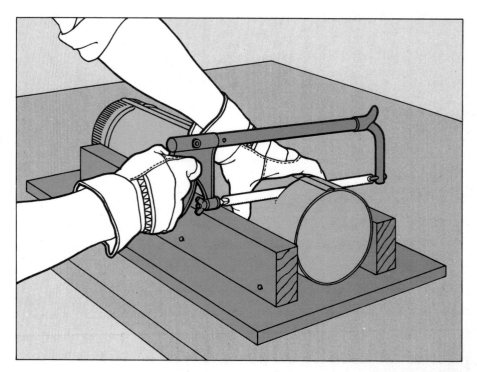

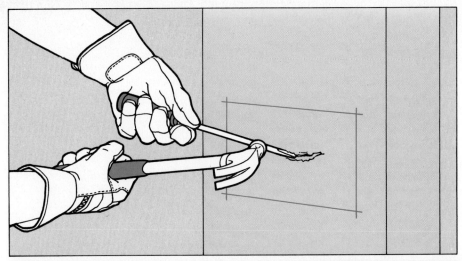

Starting a hole. Press the edge of a screwdriver blade against the duct near the center of the hole you wish to make. Hit the screwdriver shank with a hammer to make a hole large enough to admit the tips of a pair of tin snips.

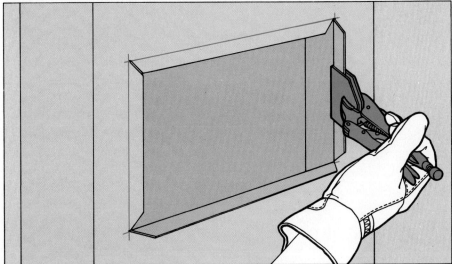

Making a flange. In ductwork, flanges are often necessary in the side or at the end of rectangular duct. (In round duct, collars serve in place of flanges.) To make a flanged opening in the side of a rectangular duct, first outline the hole dimensions on the duct. Then mark a second hole 1 inch inside the first and cut it out, using the techniques above. If the hole is rectangular, snip diagonal cuts in the corners, stopping at the outline. If the hole is circular, make evenly spaced cuts about 1 inch apart to the outline. Fold each of the tabs formed by cutting so they are perpendicular to the duct surface.

You can bend sheet metal with your hands if you wear gloves or, if you must make several flanges, use broad-billed pliers, some of which are calibrated to set the depth of the flange (*left*). To flange the end of a rectangular duct, snip 1 inch along the duct at the corners.

A Range of Registers

A register for every purpose. The three registers at right typify the models generally used to expand a heating-cooling system. All have vanes or fins to direct the flow of air into a room, and controls to reduce the flow or cut it off entirely.

To deliver air from overhead ducts, round registers with wide diffusion vanes create air turbulence, so that warm air does not build up at the ceiling and leave a pool of heavier cold air near the floor. A baseboard register is most often used on an outside wall, where it warms the cold air that flows down the windows and the wall. A rectangular register can be fitted into walls, floors or ceilings. The type illustrated is best for tapping into an existing duct that lies so close to the surface that a direct connection can be made (*pages 52-53*), as is generally the case with ducts inside walls. Ducts in the ceilings and floors may be several inches from the surface, requiring an extension box (*page 53*).

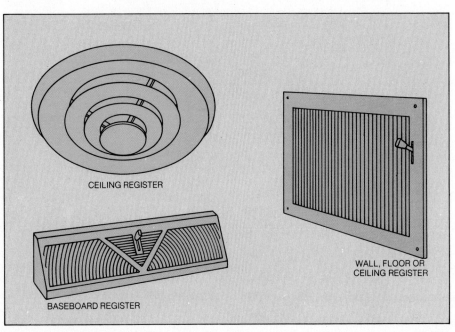

CEILING REGISTER

BASEBOARD REGISTER

WALL, FLOOR OR CEILING REGISTER

A Direct Connection for a Rectangular Register

1 **Cutting the templates.** Measure the length and width of the register front, outline these dimensions on a piece of stiff cardboard, and cut out the outlined rectangle. To allow for a collar that lies within the rear edge, or shoulder, of the register, add ½ inch to the collar dimensions and transfer these dimensions to the cardboard. Cut out this inner rectangle and save both pieces: The hollow rectangle will serve as a template to mark the surface of the wall that covers the duct; the solid one will be used as a template to mark the metal of a duct.

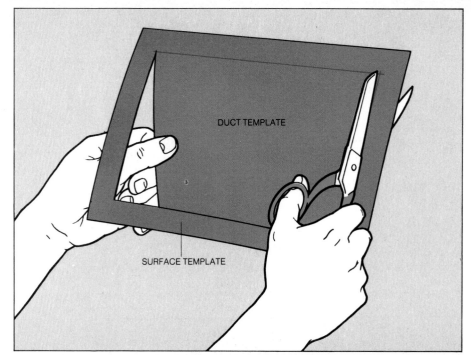

DUCT TEMPLATE

SURFACE TEMPLATE

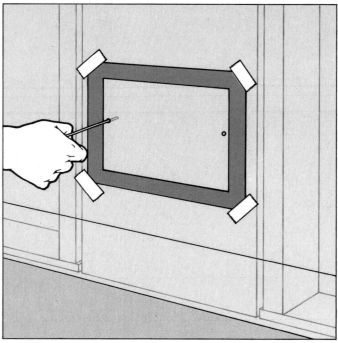

2 **Completing the duct template.** On the metal-marking template, mark lines 1⅝ inches in from each edge and snip out the corners. The long strips at the edges of the template are guides for flanges to support the register.

3 **Cutting a hole.** After you have located the duct you want to tap, set the wall template over the surface concealing it, and tape on the template. Drill ⅛-inch holes just inside the corners and probe behind the surface with a stiff wire to verify the position of the duct; if necessary, move the template to fit completely over the duct. Mark the inner outline of the template on the surface and cut out that portion with a utility knife.

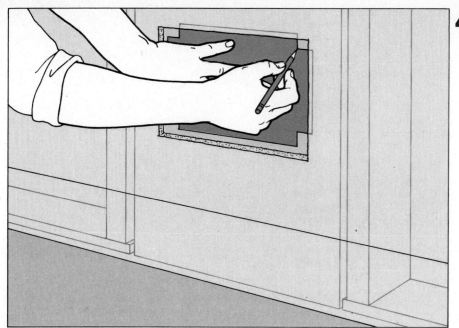

4 **Marking the duct.** Set the duct template against the duct, directly behind the opening in the wall. Mark the cutout corners of the template, then fold the template's flaps back to make a rectangle. Set the rectangle back into position against the duct, with its corners touching the points you have marked; outline the rectangle on the duct, and draw diagonal lines outward from its corners to the corners of the opening. Using these lines as guides, cut a hole and turn the flanges (*page 51*) to lie flat against the outside of the wall. Punch holes in the flanges and attach the register with sheet-metal screws.

An Extension Box for a Recessed Duct

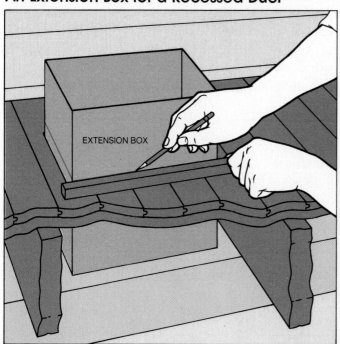

EXTENSION BOX

DUCT

1 **Cutting the box to fit.** To connect a rectangular register to a duct running deep in the joist space of a floor or ceiling, prepare templates for the front and rear of the register and make floor and duct openings by the methods shown above, Steps 1-4. Your supplier will be able to match an open-ended extension box to the collar at the rear of the register. Stand the box on the duct and mark lines 1 inch above floor level on all four sides; a scrap of lumber 1 inch thick makes a handy guide for these marks. Trim off the metal above the marks, and form a 1-inch flange on each side as shown on page 51.

2 **Installing the register.** Invert the box, slide it down into the floor opening until it touches the duct, then flatten the flanges against the duct by pushing the box down. With a long-necked awl, punch small holes through two of the flanges and the duct. Fasten the box to the duct, using sheet-metal screws and a long-necked screwdriver. Finally, set the register in place and fasten it to the floor with wood screws.

Installing Ducts and Fittings

Adding one or two duct runs to an existing system is a matter of simple carpentry and sheet-metal work. All of the parts of a duct run are standardized and easy to hook up *(pages 54-57)*. And the carpentry part of the job is simplified when you use round ducts, run outside a wall and then concealed by a wallboard-sheathed enclosure *(pages 58-59)*, rather than attempt to fit shallow rectangular ducts between the wall studs.

Round duct, called pipe, should be at least 6 inches wide: narrower pipe tends to whistle when air passes through it, and it cannot handle the large volumes of air needed for air conditioning. A combination of rigid round ducts and flexible ones, made of plastic or spring wire, will carry runs to any point in a house. For short, sharp turns, install fittings called elbows, either rigidly set at 45° or 90° or flexible for almost any angle you choose.

Your choice of a starting point for a run will depend on the pattern of your existing system *(below)*. In a radial system, runs of round duct start at the furnace plenum and radiate outward toward the rooms of the house; new runs must also take the plenum as their starting point. In an extended-plenum system, one or more large rectangular ducts run from the plenum, and ducts for individual rooms start at these main ducts; to start a run, choose a point on the extended plenum convenient to the room you plan to serve. Installation of collar fittings to start runs is illustrated opposite. The outlets, called boots, consist of short final sections, usually angled to accept a wall, floor or ceiling register.

How ducts run. No heating system resembles the one in the drawing below—but the drawing does show how new duct runs can be added to a variety of systems. In both the radial *(left)* and the extended-plenum *(right)* systems, the new runs begin with a fitting called a collar. A run may go directly to a single room *(left)* or be branched at a T or Y to serve more than one room *(right)*; in either case, each run is fitted with a damper to control flow. The elbow in the run at right is a flexible type that can be set at any angle; rigid elbows are preset at the factory. Also available are long sections of flexible duct, like the length in the run at left, which combine easy turns with a substantial run. At the ends of the runs, "boots" preset for a variety of angles protrude into the walls or the floors to receive register grilles.

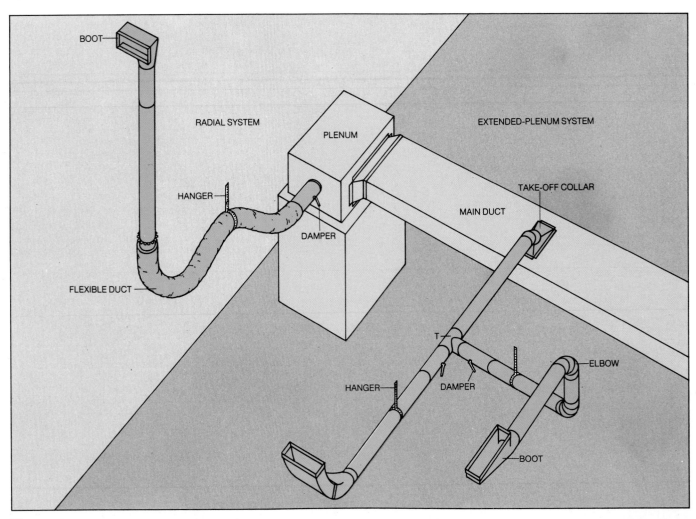

Starting a Run

A straight collar. When installing this fitting as a direct connection to a furnace plenum, wear gloves and a long-sleeved shirt. Cut a hole exactly the size of the collar *(page 51, top)* in the plenum, at least 1 inch below the top of the plenum. Slide the collar into the hole, tabbed end first, until the projecting bead around the middle of the collar meets the side of the plenum, then reach inside the collar and fold the tabs to lie flat against the inside of the plenum. The crimped outer end of the collar will connect to the plain end of the first duct section in the run.

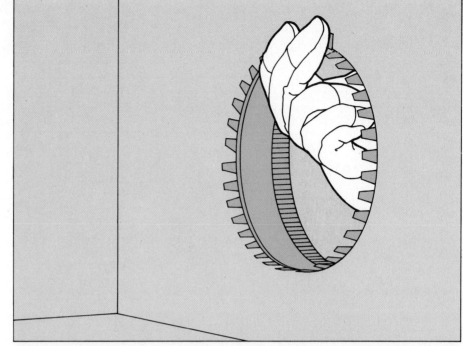

A take-off collar. A take-off is generally needed for the top or side of an extended plenum; from this point, its adjustable elbow can start a run of round duct in almost any direction, even in cramped quarters. Install a take-off—wearing gloves and a long-sleeved shirt—by cutting a rectangular hole in the plenum just big enough to admit the take-off tabs. Slide the take-off into the hole, tabbed end first, until the take-off flange rests upon the outside of the plenum; at this point the top of the elbow must clear the joists or ceiling above by at least 1 inch. Reach through the take-off and fold the tabs to lie flat against the inside of the plenum. Finally, on the outside of the plenum, drill or punch holes through two opposite sides of the flange and the plenum beneath it, and secure the take-off to the plenum with sheet-metal screws.

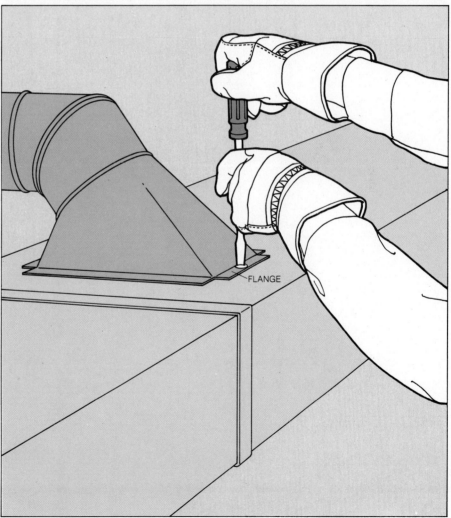

FLANGE

Assembling Duct Sections

A snap-lock assembly. Round duct is generally shipped open in nests of three to 10 sections. Before closing the seams, measure the entire duct run and shorten one section, if necessary, by snipping a length off the plain—not the crimped—end. If the snips pinch the ends flat, pry them open with a screwdriver.

The type illustrated is assembled by snapping a tongue into a slot (another type is described below). Wearing gloves, start at either end of the duct section and press the duct into a cylinder, then thread the tongue into the slot until the seam clicks shut (*inset*). Repeat the process at 1- or 2-foot intervals. Caution: Do not hammer a completed snap-lock seam for a tighter seal—you may break the seal.

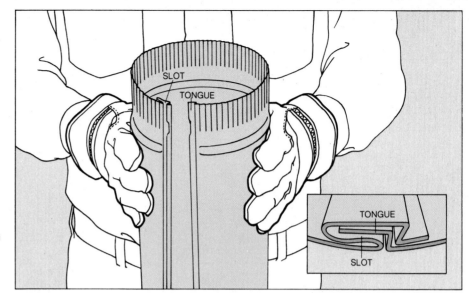

A hammer-lock assembly. In some duct designs, the seam consists of two U-shaped edges (*inset*). Shape the entire section with your hands and hook the edges lightly together. Then hang the duct over the edge of a suspended 2-by-4 and, starting at either end, hammer the seam shut. Caution: Hammer blows can easily dislodge unsealed parts of the seam. Check constantly as you hammer along the section; an error may force you to reopen an entire seam.

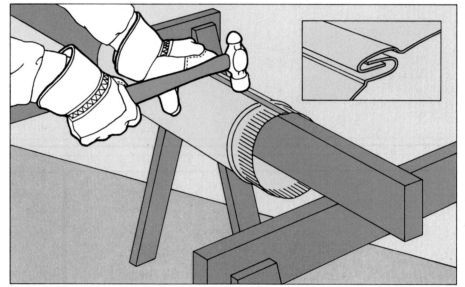

Joining a crimped to a plain end. Slide the crimped end of a duct section, set to face away from the furnace, into the plain end of the next section until the bead of the crimped section touches the edge of the plain one, then align the seams of the two sections. Just outside the bead, drill or punch small holes in opposite sides of the connected ducts, and secure the connection with sheet-metal screws. Seal the connection at the bead with duct tape.

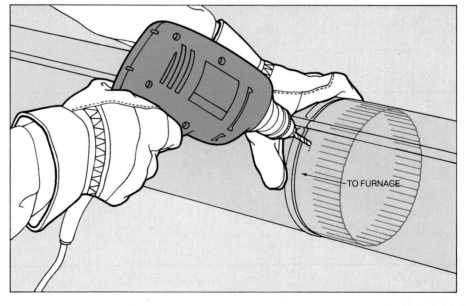

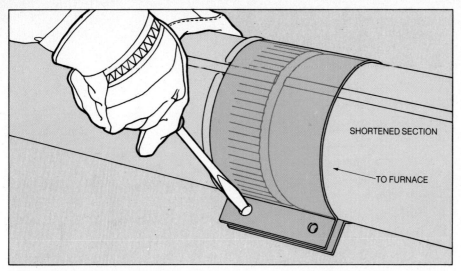

A drawband for a shortened section. To fill a gap shorter than a complete duct section, cut a section about 2 inches longer than the gap. (If the seam is already closed, cut the section at the plain end with a hacksaw, as shown on page 50.) Connect the crimped edge of the short section in the usual way *(opposite, bottom)*. The plain end should meet or nearly meet, but not overlap, the crimped end of the adjoining duct or fitting. Connect the two ends with a drawband—a flexible collar of galvanized steel, tightened by nuts and bolts. Slide the drawband along the plain duct end, bring the two duct ends together and slide the drawband back until it touches the bead of the crimped end. Tighten the drawband and seal the connection with duct tape.

SHORTENED SECTION

TO FURNACE

Dampers and Hangers

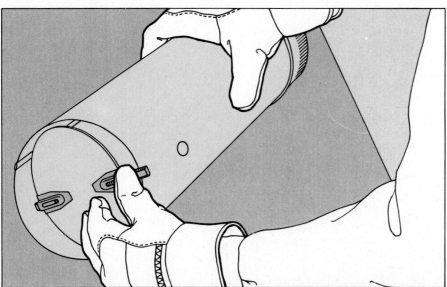

Installing a damper. Each duct run should have a damper, preferably near the furnace, to shut off heat from a room, if necessary, and to balance the entire heating system *(pages 10-11)*. Some dealers offer 2-foot duct sections with factory-installed dampers; alternatively, you can buy the damper separately and install it.

In the common model at left, the damper is held in place by spring-loaded shafts. Drill or punch holes for these shafts in opposite sides of the duct, at least 6 inches in from the plain end. Retract the shafts in their slots, set the damper in place, and release the shafts, threading them through the holes you have made. Finish by fastening the damper handle, following the manufacturer's instructions.

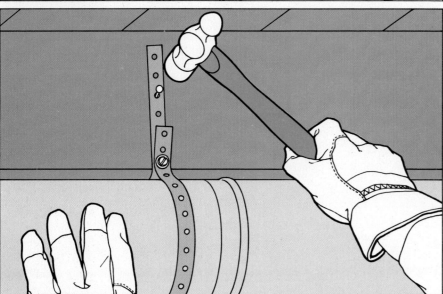

Installing a hanger. Every horizontal part of a duct run must be supported from above at intervals of no more than 10 feet; if the duct sections are relatively short—6 feet or less—shorten the intervals accordingly. The flexible hanger shown at left fits ducts of any thickness and fastens to either joists or ceilings. Secure it around the duct with a nut and bolt, then nail its other end to the supporting surface. Keep a clearance of at least 1 inch between the duct and the joist or ceiling above it.

Adding a
New Duct Run

The pictures on these pages illustrate in detail the installation of a common type of new supply duct, in which air is tapped from a furnace plenum in a basement and released high on the wall of an upstairs room. (In schematic form, this duct run appears at the left of the picture on page 54.) Such a run is especially useful in a house with newly installed air conditioning, as a complement to a heating run with an outlet low in the room (page 50). The same installation principles, however, apply to any run of new duct—for example, a heating run that ends at a floor register. In every case, you must make holes in walls, ceilings or floors to provide access to a plenum or an existing duct. After installing the new duct and connecting it to the heat source, attach a boot and a register.

You may be able to hide new duct by running it inside a closet or above a suspended ceiling. More often, you will have to conceal it in a duct enclosure. A two-sided enclosure (right) could conceal a vertical duct in the corner of a room or a horizontal duct running along a ceiling-and-wall corner. A frame with three sides could enclose a duct that runs at some distance from a corner.

Anatomy of an enclosed duct. Air passes through a rigid 6-inch round duct in one corner of a room and enters the room through a boot and a register. The duct is boxed in by two adjoining walls and by a wallboard-covered frame of 2-by-2s fastened to the adjacent walls. The boot flanges project ½ inch beyond the frame to allow for the thickness of the wallboard. The frame must be at least 8 inches wider than the boot and 3 inches deeper than its depth when measured with its flanges folded back.

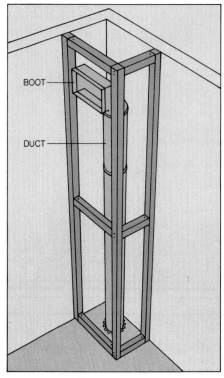

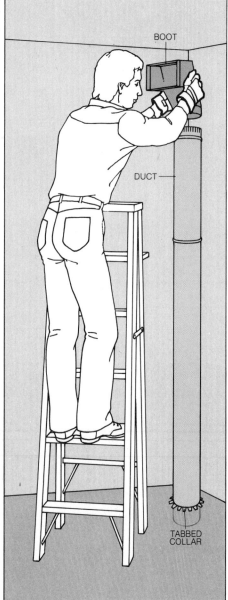

Installing the Duct

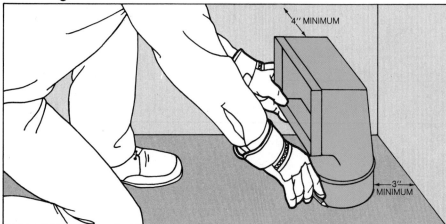

1 Positioning. Set the boot on the floor in the corner selected for the duct, facing the room's long axis; the side of the boot face should be 4 inches from the room's long wall and the back 3 inches from the short one. Trace around the base of the boot on the floor. At the center

of this circular outline, drive a nail through the floor. In the room below, check that there are no ducts, pipes, joists or other obstructions within a circle 7 inches wide around the nail. If necessary, relocate the circle upstairs until a clear passage for the duct is assured.

2 Fastening the duct in place. Using the nail in the boot circle as a center, mark on the floor a circle ½ inch wider than the boot circle and cut out the larger circle with a keyhole or saber saw. Open the seam of a starter collar (page 55, top), fold the tabs outward to a right angle, and strap the collar around the duct, crimped edge down. Let about 8 inches of the plain end of the duct protrude from the collar. Slide collar and duct into the hole and nail every third tab to the floor. About 5 or 6 inches of duct should show in the room beneath.

Mount the boot at the top of the duct. The boot's base should be 6 feet above the floor. If necessary, lengthen the duct with a piece from a 2-foot section, sold as a flat sheet. Cut a sheet to length at the plain end, then snap the sides together to make a round insert.

3 **Connecting to the plenum.** Cut a hole in the sheet-metal plenum of the furnace and install a tabbed starter collar in the opening *(page 55)*. Over the crimped end of the collar fit the plain end of a 2-foot section of pipe containing a damper. Fit the plain end of a flexible duct over the other end of the damper section and insert the crimped end of the duct into the plain end of the rigid duct protruding into the basement; if the basement is unheated, use flexible duct with factory-installed insulation and vapor barrier. Seal all joints with duct tape and support the damper section and the flexible duct with duct hangers.

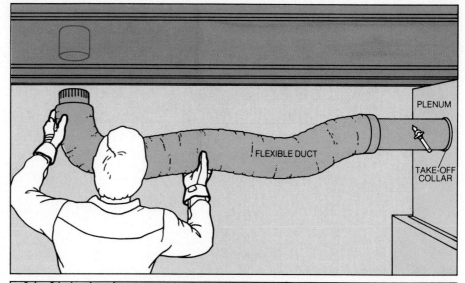

PLENUM

FLEXIBLE DUCT

TAKE-OFF COLLAR

4 **Constructing the frame.** Determine the dimensions of the frame and cut the pieces. Lay out the parts of the wider of the frame's two walls on the floor. Space the crosspieces 3 or 4 feet apart, adjusting the spacing to leave room for the register between the two top crosspieces. Nail the uprights to the crosspieces with eightpenny nails. Temporarily toenail the crosspieces for the narrower side to the corner upright with finishing nails *(right)*, and attach the third upright to these crosspieces with eightpenny nails. Brace the assembly against a wall and secure the toenailed joints with eightpenny nails.

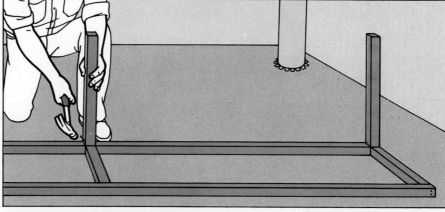

5 **Installing the frame.** Cut out sections of baseboard, if necessary, to accommodate the finished enclosure. Set the frame in position with the boot flanges just outside the enclosure. If necessary, adjust the duct angle to align the boot properly. Make sure the uprights of the frame are vertical and the braces horizontal, if necessary inserting thin wood wedges, or shims, between the frame and the wall. Nail the uprights to the wall studs with 16-penny nails, or drill holes for wallboard anchors.

WALLBOARD   BOOT

6 **Finishing the enclosure.** Cut wallboard to fit the wide side of the frame. Set it against the frame and reach inside to mark the positions of the boot flanges. Lay the board on the floor, remove the boot and set it over the outline. Reaching inside the boot, outline the register hole. Cut the hole out of the wallboard. Replace the boot and install the wallboard. Cover the narrow side of the frame with wallboard, cement and tape the seams, and install baseboard. Install the register.

Space Heaters to Warm Individual Rooms

If tapping the existing system for extra heat turns out to be impractical—because the furnace has little reserve, or the connections are complicated to make—you may be able to get the additional warmth you need from a space heater: one of the small units made to heat small spaces. They may consume wood or coal *(pages 82-85),* gas or oil *(pages 62-65),* electricity *(pages 72-77)* or no fuel at all *(below and right).* This last type, which is called a flue heat exchanger, extracts ordinarily wasted warmth from the exhaust gases of the furnace.

For spaces in or near the existing furnace room, the flue heat exchanger is particularly useful, since it takes advantage of the flue already installed there. It is practical, however, only if the system meets certain conditions. Since the exchanger reduces draft in the flue as it extracts heat, you must be sure that the temperature and draft are both sufficient to sustain this loss without impairing furnace operation. The only way to be certain is to test *(page 31, Step 3);* only if the temperature is between 500° and 800° and the draft measures at least .04 inch of water *(page 30, Step 2)* should you consider a heat exchanger. Also, if you duct heat out of the furnace room, enough fresh air must enter it so the exchanger blower does not starve the furnace of air needed for combustion.

To provide room for attaching an exchanger, there must be a straight section of flue pipe at least 1 foot long, and it must offer safe separation from combustible materials such as flooring and joists. Check your local code. If you have a gas burner it makes no difference where on the flue this section is, but on an oil burner the straight section of flue must be between the stack control and draft regulator *(page 30, Step 2).* To get the space, you can move the stack control closer to the furnace *(opposite, top)* or switch sections of flue to move the draft regulator closer to the chimney. Make sure that the regulator flap is vertical and that the hinge is horizontal. After a heat exchanger is installed, the furnace should be adjusted as explained on pages 30-33 to maintain an adequate draft.

Two heat exchangers. In one type *(right, top),* which replaces a section of the flue, furnace exhaust gases pass through a battery of pipes, heating them and activating a thermostat that turns on a blower. The blower circulates room air between the hot pipes where it is warmed before leaving the unit through a warm-air outlet.

The second kind of heat exchanger *(below)* uses heat collectors inserted into the flue. Water in the collectors transfers heat from inside the flue to the outside. A thermostatically controlled blower circulates air between the pipes to heat it. Both units are sealed to prevent flue gases from mixing with room air.

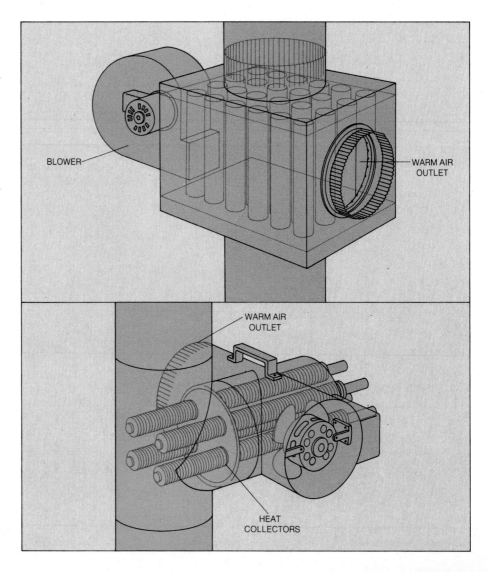

BLOWER

WARM AIR OUTLET

WARM AIR OUTLET

HEAT COLLECTORS

Moving a stack control. If the stack control must be relocated, turn off the furnace at the emergency switch and the service panel. Remove the stack control (*page 27, bottom right*), then unscrew the stack-control bracket. Mark a location nearer the furnace for a new stack-control hole, making sure that the control will fit between the new hole and furnace and that the wire on the control will reach. Start the hole with a drill and enlarge it with tin snips to the size of the old hole. Place the stack-control bracket over the new hole and mark the stack for screw holes. Drill the holes, attach the mounting bracket, then remount the stack control (*page 27*). If the old hole in the flue section is not removed when the heat exchanger is installed, patch it with sheet metal fastened with sheet-metal screws. The patch need not be airtight.

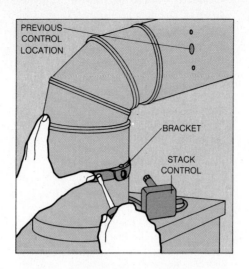

PREVIOUS CONTROL LOCATION

BRACKET

STACK CONTROL

A Heat Exchanger Fitted into a Flue

1 Trimming the flue. Be sure the furnace is off. Wear protective gloves. Remove a section of flue by unfastening any screws that secure the joints and then pulling the joints apart, removing the flue from the breech of the furnace if necessary. Measure the length of the heat exchanger from the end of the plain flue collar to the start of the crimping on the crimped flue collar. From the plain end of the flue section you removed, cut a piece equal to the length of the heat exchanger, as shown on page 50. Discard this piece, then fit the remaining crimped piece of flue into the heat exchanger (*right*). Drill three ⅜₂-inch screw holes and fasten the flue piece to the heat exchanger, using ¼-inch, No. 6 sheet-metal screws.

2 Installing the heat exchanger. Enlist a helper to fit the heat exchanger into the flue. New screw holes may be necessary to fasten the flue to the heat exchanger. Reattach the flue to the breech of the furnace if you have loosened it. Use stovepipe wire wrapped several turns around nails in joists to support the heat exchanger at both ends. Restore power to the furnace and plug in the heat exchanger.

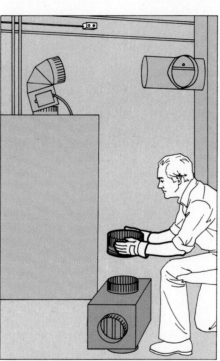

A Heat Exchanger Fitted to a T

CROSS PIECE

STEM

Angling the heat collectors. Wearing gloves, replace either a vertical or a horizontal section of flue with a T that fits both the flue and the heat exchanger. If necessary use the techniques in Step 1 and on page 50 to shorten the plain end of the T crosspiece to fit the flue and to trim the T stem so that it projects 1½ inches from the crosspiece. Install the T in the flue with ¼-inch, No. 6 sheet-metal screws. Have a helper push the heat exchanger into the stem as far as it will go and then hold it in place. Drill a ⅜₂-inch hole through the top of the T stem and heat exchanger, being careful not to drill into the heat collectors.

Using ¼-inch sheet-metal screws, attach the heat exchanger to the top of the T stem. Have your helper raise the outer ends of the heat collectors at least ½ inch higher than the ends inside the flue (*above*) while you drill a hole through the bottom of the T stem and heat exchanger and screw them together. If you install the unit in a horizontal flue, put it on top so the heat collectors point down.

A Heater Hidden in a Wall

The traditional space heater, an awkward metal box, stands away from the wall and has a thick black pipe coming out of its back. Similar heaters are still available *(page 65)*, but for the most part, free-standing models have given way to thermostatically controlled gas "furnaces" that can be mounted on a wall, with their exhausts vented through camouflaged or completely invisible pipes.

The easiest of these wall furnaces to install is the direct-vented type that is illustrated on these pages, in which two concentric pipes go directly through an outside wall from the back of the heater. The outer pipe brings air into the heater to support combustion; the inner one carries exhaust fumes out of the house. An up-vent unit that releases its exhaust through the roof can be installed on any wall directly below an attic *(page 64)*. The flue for an up-vent heater must rise at least 12 feet above the bottom of the unit, and should rise as near to the vertical as possible—angles and horizontal runs reduce the draft. In either design, choose a model that draws air from the top of the room, heats it and blows it out near the floor—such a "counterflow" unit keeps a room more evenly heated from floor to ceiling.

Your local building code may restrict your choice of a model and your freedom to install one. Some codes require electric ignition rather than a pilot light, some forbid a homeowner to make his own installation, and at various times of fuel shortages, a few areas have declared moratoriums on all new gas connections.

For your own part, be sure that the addition of a space heater will not require a costly enlargement of the gas supply line to your house. And though a wall furnace should be placed as near to the middle of a wall as possible, be sure that it is in a spot where gas and electricity can be supplied without difficulty and where the unit can be correctly and easily vented. Direct-vent units must clear the ground by at least 1 foot and the corners and overhead projections of the house by at least 2 feet. An up-vent unit, which requires an attic above the space to be heated, must have the path for its vent cleared of flooring and insulation. And within the room in which any unit is installed, you must maintain clearances—usually specified by the manufacturer—from furniture, doors and drapes.

Unless you have had extensive experience in assembling and testing piping for gas, have a professional make the gas connections. But you can make the electrical connections for a power supply and a thermostat yourself. Many heaters are simply plugged into a receptacle. Some heaters come with a separate thermostat, and with wires for the thermostat connections already attached to the unit. Choose a location on an inside wall for the thermostat *(page 16)*, run the wires from the furnace to the wall *(pages 72-75)* and mount the thermostat *(page 16)*.

1 **Hanging the heater.** Locate two adjoining studs by drilling small test holes through the wall and probing with stiff wire. Midway between the studs saw a vent hole of the size and height specified for your unit. Remove any insulation from inside the hole, then drill through the outside wall at the center of the hole. In the outside wall, cut a vent hole the same size and ¼ inch lower than the one on the inside. Use a saber saw or keyhole saw to cut these holes, unless the wall is masonry, which must be chiseled through from the outside. Using the inside hole and the back of the heater as guides, find the positions for the wall brackets that support the unit and screw these brackets to the wall. Locate a position in the floor for the gas inlet pipe and drill a hole for it. With helpers lifting the unit, guide the vent-pipe collar into the vent hole and slip the mounting slots over the wall brackets.

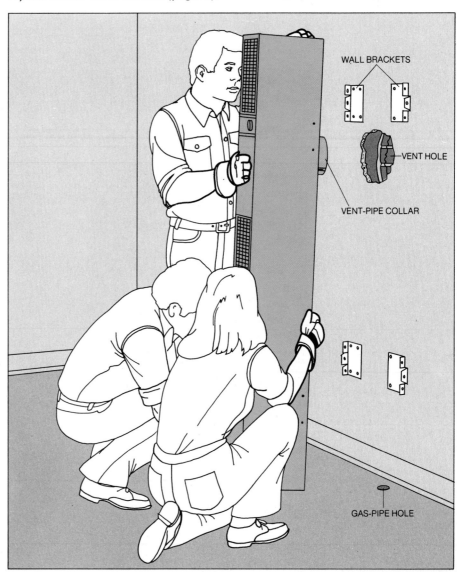

WALL BRACKETS

VENT HOLE

VENT-PIPE COLLAR

GAS-PIPE HOLE

2 **Installing the intake pipe.** From outside the house, fit the outer pipe of the vent over the furnace collar. If the pipe has telescoping sections, adjust them so that it protrudes from the wall the distance specified by the manufacturer—typically, ½ inch. Otherwise cut the pipe to the proper length with a hacksaw *(page 50)*. Set the pipe back on the furnace collar, positioning it to slope slightly downward to the outside. Check the slope with a carpenter's level and, if necessary, enlarge the hole in the outside wall to maintain the required clearance.

An air-inlet screen and an outer wall plate come as a single assembly. For a nonmasonry wall, spread caulking over the back of the wall plate, position the entire assembly over the vent pipe and fasten the plate to the wall with screws. For a masonry wall, position the assembly and drill holes through the plate into mortar joints with a small masonry bit. Remove the plate and enlarge the holes with a larger bit, to accept screw anchor fasteners, then caulk the plate and screw it to the wall. In both surfaces, finish the job by caulking the plate-to-wall seam.

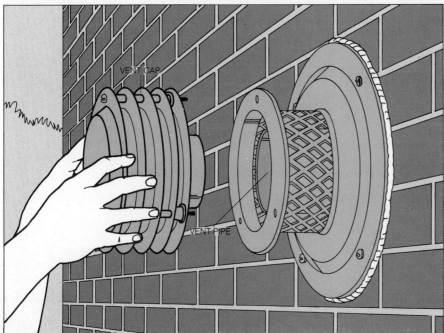

3 **Installing the vent cap.** Fit the vent pipe over the furnace flue collar, measure its length outside the wall, and cut it to the length specified by the manufacturer—typically, 3½ inches beyond the wall. Replace this pipe, fit the vent cap over it and fasten the cap to the wall plate with the nuts and bolts that are provided with the unit.

An Up-Vent Unit

1 **Making a vent opening.** Drive a nail through the ceiling next to the wall you have chosen for the heater. In the attic, using the nail as a location guide, cut a 14½-by-4½-inch hole between two joists through the ceiling below. Form a square box around the hole by nailing two blocks of wood between the joists, 14½ inches apart; use wood blocks that match the dimensions of the joists—usually 2 by 6. Remove all insulation from inside the box.

Bend the nailing tabs of a Type B-W vent fire-stop spacer inward to fit the box and nail the fire-stop spacer directly over the hole, with its top projecting ¼ inch above the tops of the joists. (The drawing below shows a spacer nailed to the joists; if the joists run parallel to the wall on which the heater is installed, nail the spacer to the wood blocks.) Close the top of the box with two rectangles of ⅜-inch wallboard of the fire-resistant type called Type X, cut to fit and nailed to the tops of the joists or wood blocks.

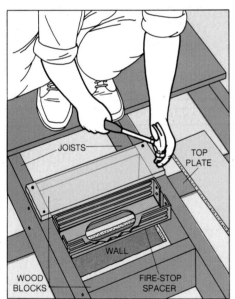

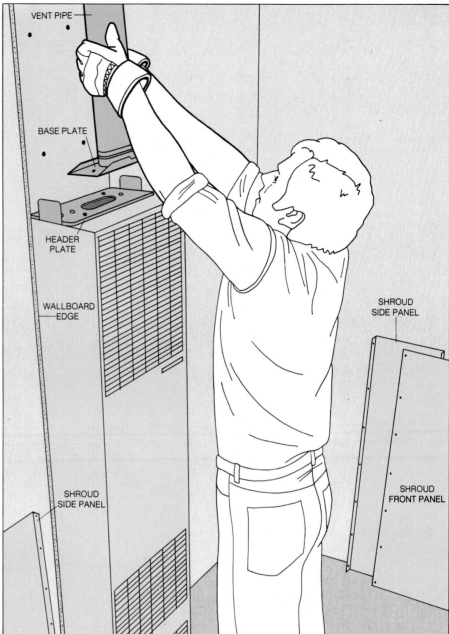

2 **Completing the installation.** Cut a rectangle of ⅜-inch Type X wallboard as wide as the furnace and as high as the ceiling. Fasten this rectangle to the wall directly beneath the vent opening. Make a hole in the floor for a gas-supply pipe; follow the manufacturer's instructions to locate this hole precisely. Set the furnace in place and attach the header plate, which secures the vent pipe to the top of the unit. Install a Type B-W vent pipe long enough to rise at least 2 inches above the bottom of the fire-stop spacer; fasten the bottom of the pipe to the vent base plate, angle the pipe up through the spacer, and screw the base plate to the header plate. If necessary, tip the top of the furnace out of the way to allow the vent pipe to clear it. In the attic, install a Type B-W to Type B 4-inch round adapter at the top of the vent pipe, then finish the vent-and-chimney assembly (*pages 84-85*).

Secure the unit to the wall with toggle bolts or screws or, in some models, with brackets mounted near the header plate. Run No. 14 electric cable to the junction box of a heater that does not have a plug-in cord; staple the cable to the wall at least 2 inches away from the vent pipe, but within the part of the wall that will be concealed by the vent-pipe shroud. Cut the shroud panels to fit between the furnace and the ceiling. Fasten the side panels to the wall with toggle bolts or screws, then attach the front panel to the side panels. You will be left with two unfinished wallboard edges behind the heater; they can be concealed with wood molding.

Freestanding Space Heaters

Freestanding heaters may burn any of a variety of fuels—oil, gas, coal or wood. Except for the gas-fired type, freestanding heaters require an "all fuel" vent pipe; a gas heater should be vented through a Type B pipe. Check your local building code for any safety devices required by law and for the required minimum clearance from any combustible materials, and plan your installation for the straightest, simplest fuel and vent paths. When the installation is complete, check it with a draft gauge *(page 30, Steps 1 and 2)*; if the draft cannot be adjusted to meet the manufacturer's specifications *(page 30, Step 3)*, then you must either lengthen the chimney or else install a draft blower.

Because fuel spills are very hazardous, have a professional install the line unless you are sure you can make absolutely leak-free connections. Never use a torch to repair connections.

An oil heater. Set on a protective stoveboard of metal and asbestos fiber, this heater is screwed to the floor at a location that provides generous clearance from combustible materials. Single-wall smoke pipe is attached at the ceiling to a Type A ceiling support component *(page 85, Step 1)* and contains a barometric draft regulator installed in a T. An oil line runs to the heater from an outdoor tank containing No. 1 fuel oil; the bottom of the tank must be at least 6 inches higher than the heater control valve, and the top less than 12 feet above this valve. The tank itself is set on a concrete slab or solid blocks, and tilted toward its outlet valve. This valve and a trap allow condensation to be drained off from the tank. A filter protects the heater from sludge, and a shutoff valve at the heater stops the fuel flow during repairs.

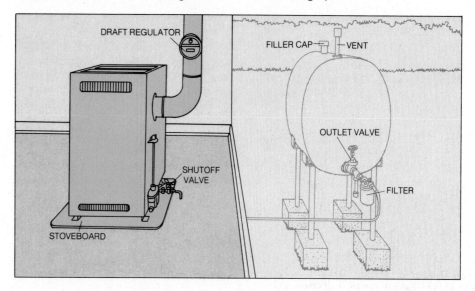

Making the Connections

1 **At the oil tank.** Apply a thin coat of pipe-joint compound made especially for oil-pipe connections to the male threads of pipe fittings. Reduce a ½-inch tank-outlet fitting to ⅜ inch with an adapter, and fasten a short nipple and a valve to the adapter. Assemble a trap from a 4-inch nipple, a cap and a T; fasten the trap to the valve with another short nipple. Attach an oil filter to the trap with a third short nipple, and fasten a male flare fitting to the filter. Run copper tubing from the oil tank to the heater, and fasten it to the filter with a flare nut.

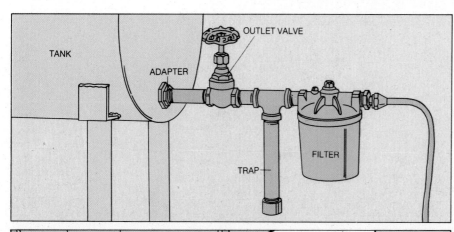

2 **At the heater.** Install a flare fitting on the inlet hole of the oil heater's control valve and attach a short length of tubing, cut with a tube cutter, to the fitting. Attach a valve with a flare fitting to this tube and to the tube from the oil tank. Caution: Put some oil in the tank and check to make sure all connections inside and out are leak-free. Install a chimney *(pages 82-85)*, and run a length of smoke pipe containing a T joint from the heater to the chimney. Install the barometric draft regulator in the stem of the T, with the hinge horizontal and the flap vertical. Adjust the regulator *(page 30, Step 3)*. Level the furnace across the front and from front to back, using the adjusting screws at the base. Set the level on top of the oil control valve as shown; alternatively, set it on top of the heater.

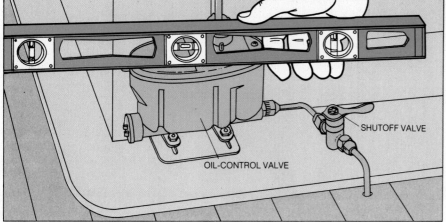

Adding a Convector to a Hot-Water System

Most hot-water systems have reserve capacity enough to handle one or even two additional upright or baseboard convectors—and installing them is a basically simple job of tapping into the supply or return mains and running two pipes as the supply and return riser connections to the convectors.

Before you start installing a new convector, find the section of the supply line closest to the place where you want to put the fixture—preferably on an exterior wall and under a window. The supply line may follow the perimeter of the out-

er walls under the floor joists (or above the joists on a slab foundation), or it may run along the center beam of the basement or crawl space with the risers between the joists. Once you locate the main, check the water flow *(below)* so you can put an inlet valve on the incoming or supply riser and a bleeder valve on the outgoing or return riser. If the fixture is directly above the main, you can run the risers straight. But in most cases you will make a zigzag or dog-leg for each riser with two elbows *(below)* to connect the convector to the main.

In a typical system, the mains are 1 inch in diameter, the risers are ¾ or ½ inch. When measuring the pipe lengths, make sure to allow for the extra ½ inch or so at each end that will be sweated or threaded into a fitting.

Most existing home piping is steel, but most convector and baseboard pipes are copper. You usually can use either for new pipes, but copper is easier to work with. Where local codes ban copper pipe because minerals in the water supply will clog it, the risers will have to be made of steel joined with threaded fittings.

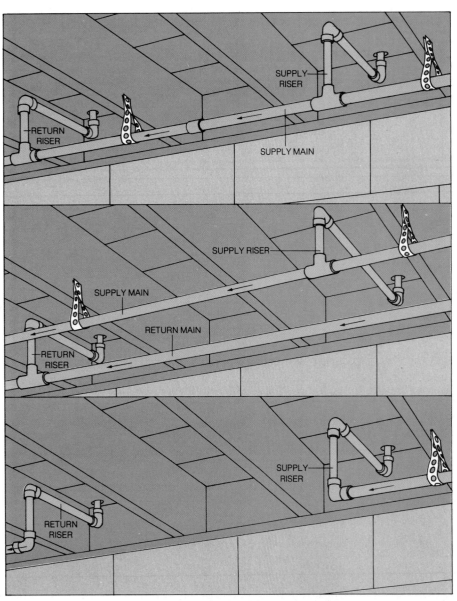

Checking the supply line. Find the section of the supply main nearest the location for the new convector. Determine the direction in which the water flows by tracing the line back to the boiler. The supply end of the main leads out of the boiler at a higher level than the return end.

Now inspect an existing radiator or convector connection to see what fittings you need for tying into the line. For a one-pipe system *(top)* such as that shown on page 12, use a regular T for the supply riser where water enters the convector and a Venturi T—which maintains an even flow of water through the convector—for the return riser where water leaves. For a two-pipe system *(center)*, use a regular T to join each riser to the separate supply and return mains. For a series loop system *(bottom)*, use elbow joints to connect both risers to the main and make the fixture an integral part of the hot-water flow pattern.

(Image labels: SUPPLY RISER, RETURN RISER, SUPPLY MAIN, SUPPLY RISER, SUPPLY MAIN, RETURN MAIN, RETURN RISER, SUPPLY RISER, RETURN RISER)

Connections for a Copper Main

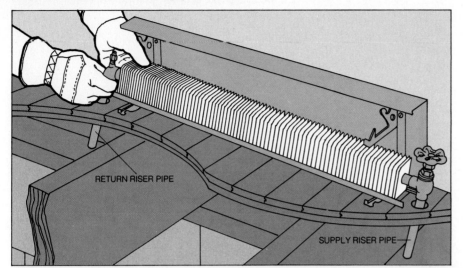

RETURN RISER PIPE

SUPPLY RISER PIPE

1 **Preparing the convector.** Drain the system *(page 42, Step 1)*. For a baseboard unit on a one- or two-pipe system fit on—but do not sweat—an elbow with a vent tapping and an angle valve. Drill riser holes and mount the back panel and heating element, using the techniques on page 47. With a series loop system, use plain 90° elbows. For an upright, screw on transition fittings and drill riser holes *(page 46, Step 2)*.

Measure from the bottom of the convector fittings to 6 or 7 inches below the basement ceiling. Cut two copper pipes this length. Remove the fittings from the convector and sweat the pipes to them. For a baseboard convector, screw on a bleeder valve and replace the disk assembly. Put the elbows or transition fittings back onto the convector, inserting the pipes into the riser holes.

RETURN RISER PIPE

SUPPLY MAIN

SUPPLY RISER PIPE

2 **Cutting the main.** Line up a straightedge between the bottom center of the return riser pipe and the main, and mark where the straightedge crosses the main. For a one-pipe system, draw a line outside the mark at a distance equal to the measurement from the stop point or bulge at the plain end of a Venturi T to the center of the stem. Repeat the same procedure for the supply riser, using a regular T. Use a hacksaw or tubing cutter to cut the main at both lines.

For a two-pipe system, mark both the return and supply mains, and measure a regular T to draw both cutting lines. For a series loop, measure a 90° elbow from one opening to the stop point level of the other opening to draw the lines.

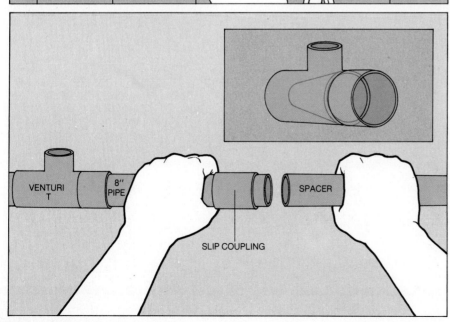

VENTURI T

8" PIPE

SPACER

SLIP COUPLING

3 **Tying into the main.** For a one-pipe system *(left)*, fit the Venturi T—with the flared mouth of the inner pipe facing inward toward the supply riser *(inset)*—onto the outgoing end of the main. The inner pipe of the Venturi creates enough suction to keep hot water moving through both the main and the convector. Fit the plain T onto the incoming end of the main. Cut an 8-inch piece from the pipe section you removed, slide a slip coupling on it and fit one end of the pipe into one of the Ts. Then cut a spacer to fill the remaining opening in the main. Gently pull one end of the main to the side and fit the spacer into the second T. Release the main so the spacer meets the end of the 8-inch pipe. Then slide the coupling over the joint. Fit a pipe 2 or 3 inches long into the top of each T and use a level to be sure the Ts are plumb. Sweat all the joints.

For a two-pipe system, follow the same procedure, but use regular Ts for both risers. For a series loop, substitute 90° elbows for Ts.

4 **Installing the risers.** To determine the length for the horizontal sections of each riser, measure from the center of the opening of the 3-inch pipe to the center of the opening on the adjacent existing riser pipe. Deduct for two 90° elbows. Cut horizontal sections and fit two opposite-facing elbows onto them. (With an upright convector, add a shutoff valve (*page 44*) to the supply riser section.) To complete each riser, set one elbow on the pipe at the main and swing the horizontal section over to the existing riser. Mark the place where the stop point at the top of the second elbow meets the riser. Remove the riser, then cut it at the marking and fit it back onto the convector. Slip the cut end into the elbow on the horizontal section. Sweat all the joints.

For an upright convector on a one- or two-pipe system, insert a bleeder valve into the convector pipe at the return end and replace the disk assembly of the shutoff valve. For a series loop system you do not need valves at the convector. Refill the system (*page 42, Step 2*).

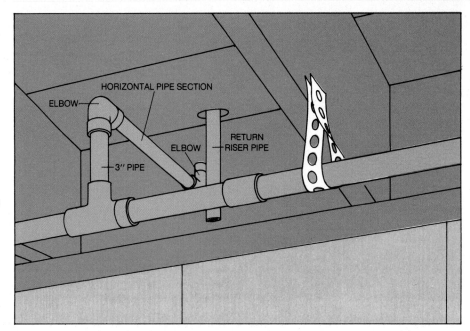

Connections for a Steel Main

1 **Measuring the main.** Prepare the convector, fit risers to the convector and mark the main (*pages 67, Steps 1 and 2*). To plan the steel replacement section for a one-pipe system (*left*), first measure along the incoming end from the nearest existing fitting beyond the mark for the supply riser. Deduct for a union, a nipple and one end of a regular T. Measure the distance between the riser marks and deduct for one end of a regular T and of a Venturi T. Then measure from the return-riser mark to the nearest fitting beyond it on the outgoing end of the main.

For a two-pipe system, measure from the marks to the existing fittings at both ends and deduct for a union, a T and a nipple on the incoming end of both mains. For a series loop system, make only the incoming and outgoing end measurements; you will not need unions or nipples.

2 **Cutting into the main.** Use a hacksaw to cut into the main between the riser marks, then unscrew the cut pieces of pipe from the fittings at both ends of the section. If the convector spans a fitting, you will have to cut and unscrew both sections of the main.

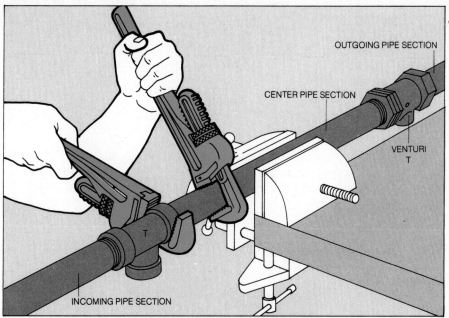

3 **Assembling the replacement section.** For a one-pipe system (*left*), screw the pipe for the outgoing end of the section onto the plain end of the Venturi T. Then attach the center pipe to the end of the Venturi T where the flared mouth of the inner pipe is visible. Add a regular T to the center pipe and finish with the pipe for the incoming end. Finally, attach one union nut of a union fitting to the free end of the incoming pipe.

For a two-pipe system, make up two new pipe sections using a regular T in both. Attach a union nut to the free end of the incoming pipe of each section. For a series loop system, attach an elbow to each of the pipes.

OUTGOING PIPE SECTION

CENTER PIPE SECTION

VENTURI T

INCOMING PIPE SECTION

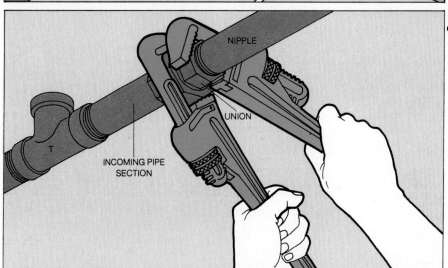

4 **Installing the replacement.** For a one-pipe system (*left*), attach a nipple to the existing fitting at the incoming end of the supply main. Slide the ring nut of the union over the nipple and screw the remaining union nut to the free end of the nipple. With a helper, set the assembled pipe section in place. Keeping both Ts vertical, attach the outgoing end of the assembly to the existing fitting at the outgoing end of the main. Fit the union nuts at the incoming end together, slide the ring nut over them and screw it tight.

For a two-pipe system, attach nipples to the incoming end of each main and use unions to connect the nipples to the pipe assemblies. For a series loop system, simply attach the incoming and outgoing sections of prepared pipe.

NIPPLE

UNION

INCOMING PIPE SECTION

5 **Connecting the risers.** Attach a steel-to-copper transition fitting to each T or elbow. Then fit a pipe 2 or 3 inches long into each transition fitting and install the risers (*page 68, Step 4*). On a one- or two-pipe system, install a bleeder valve on the return inlet and replace the valve disk assembly on the supply inlet or riser. On a series loop system, no valves are needed on individual convectors. Refill the system (*page 42, Step 2*).

TRANSITION FITTING

Stealing Heat to Warm a Room

The ingenious device shown below can heat a room by stealing hot water from any available source and, using a water pump and fan powered by a single motor, converting the hot-water heat to warm air. The adaptable unit, called a fan-coil heater, is available in models that can be flush-mounted against a wall, installed in the kick space under a cabinet, or recessed between wall studs *(inset, below)*. The source of hot water can be a hot-water boiler, a steam boiler, or, if local codes permit, a domestic hot-water tank. Except when used with a steam boiler, when it must be located in the basement *(top, right)*, the heater may be installed in any room in the house where water pressure is sufficient to keep the heater constantly filled with water.

Of course hot-water heat is not free—the BTUs removed by the heater must be restored to the boiler or tank. But the fan and pump, which use little more electricity than a 100-watt bulb, can efficiently extract heat even from relatively low-temperature water. Wall-mounted mod-

els can provide up to 8,000 BTUH from 140° domestic hot water and as much as 16,000 BTUH from 200° boiler water.

Apart from two simple electrical connections *(bottom, right)*, the main installation work consists of running ¾- or ½-inch copper tubing from the hot-water source to connections on the heater. Most boilers have plugged openings, called tappings, for alternate placement of pipes or fittings. By removing the threaded plugs and using reducing and transition fittings, these tappings can be used to connect the tubing. When direct connections to the boiler or tank are not possible, Ts can be used to connect the heater to existing hot-water piping. In some situations, a boiler drain opening can double as a connection point for the heater's return line *(top, right)*.

Before beginning any work on a boiler or tank, shut off the electrical and cold-water supplies and completely drain the unit. Install gate valves on both supply and return lines to the heater so it can be shut off in summer or for servicing.

How a fan-coil heater works. When installed between wall studs and plugged into an electrical outlet *(inset)*, only the heater's air grilles and control panel are exposed. The unit's working parts are revealed in the cutaway drawing at right. The motor, controlled by an automatic thermostat, simultaneously rotates both the fan and a drive magnet attached to the end of the motor shaft. The drive magnet is centered over a second magnet attached to the pump impeller. The rotating drive magnet causes the impeller to turn, pushing hot water—from any source—into the finned heating coils above. The fan circulates room air over the coils. A bleed valve, located on the inlet side of the heating coil, will remove air from the system before start-up and at the beginning of each heating season.

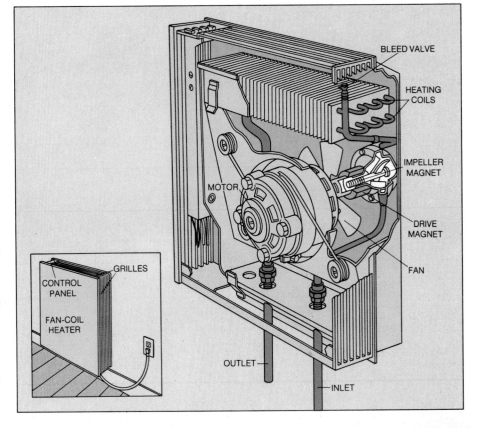

Tapping hot water from a steam boiler. Both the supply and return lines must be connected to the boiler below the normal water level, which is indicated by the middle of the gauge glass. The supply line is connected through an unused tapping to the upper portion of the boiler, where the water is hottest, but should be at least 4 inches below the water level so that steam will not enter the heater lines if the water level fluctuates. The return line can be connected to a tapping or a drain opening (*right*) in the lower part of the boiler. If you use the drain opening, install the drain cock in a T as shown; the boiler can then be drained by closing the gate valve on the heater return line and opening the drain cock. The heater itself must also be installed below the boiler water level; unlike a hot-water boiler or tank, a steam boiler will not keep the lines to an upstairs heater filled with water above the boiler water level.

Tapping a hot-water boiler. On a gravity hot-water system, the heater can be connected to the boiler, as in a steam installation (*above*), or to convenient hot-water piping. On a forced hot-water system (*right*), connections to existing piping permit the heater to operate independently of the circulator. The heater supply line is connected to the supply main on the boiler side of the flow-control valve, which opens only when the circulator is running. The return line is connected with a T and nipple through a drain opening on the boiler side of the circulator.

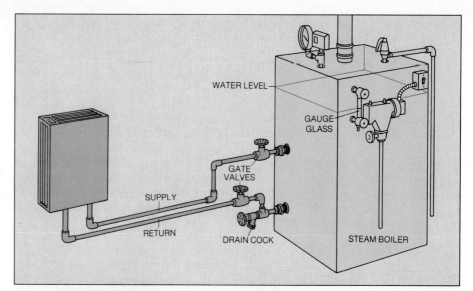

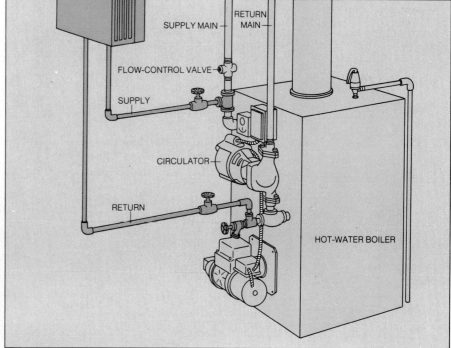

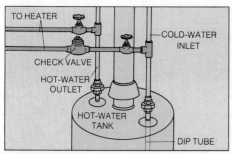

Tapping heat from a hot-water tank. Use a T to connect the heater's supply line to the existing hot-water outlet just above the tank (*above*), or at the hot-water pipe most convenient to the heater. The heater's return line must be run back to the tank and connected to the cold-water inlet with a T; the dip tube inside the tank channels the cooler water to the lower part of the tank for reheating. A check valve in the heater's return line prevents reverse circulation of cold water to the heater when the pump is not running; gate valves on both lines permit the unit to be isolated between heating seasons.

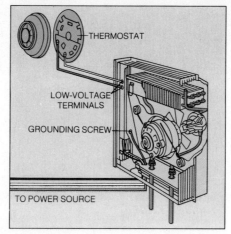

Wiring the fan-coil heater. Since the maximum current drawn by the motor is only 1½ amperes, the heater can be plugged into a convenient outlet. Use a cable or power cord with a ground wire, which should be fastened to the grounding screw on the heater frame. At the heater's high-voltage junction box, connect the hot wire to the two black wires and the neutral wire to the two white wires. Mount the 24-volt thermostat, which is supplied with the unit, and run No. 18 two-wire cable from the thermostat terminals to the two low-voltage terminals at the heater (either wire can go to either terminal).

Electric Heaters: Quiet, Clean and Durable

Electric heaters produce no smoke, fumes or gases and require no vents. They are quiet, durable and easy to maintain, requiring only an occasional vacuum cleaning to ensure maximum efficiency. These advantages make them a popular choice wherever extra heat is needed, whether the amount required is small or large.

The simplest type is the baseboard heater *(pages 76-77)* available in units 28 to 120 inches long but only 7½ inches high and 3 inches deep. Air warmed by the heater rises and gradually warms the room. Kick-space heaters *(page 77)* that are set into the spaces beneath cabinets, and heaters that are attached to or recessed into walls *(below)*, have built-in fans to force air over the heating element and into the room.

Ceiling heaters are of two types. A radiant panel can be substituted for an existing panel in a suspended ceiling. A ceiling heat lamp *(pages 75-76)* can be hung from or recessed into a ceiling. Both types provide instant heat but are highly selective, warming only directly beneath them. If a panel or a lamp above the space in front of a workbench is set to warm a person working at the bench, it will not warm the bench itself.

To determine the type and size of heater you need (measured in watts) first use the appendix *(pages 124-125)* to arrive at an estimate in BTUH and convert the result to watts *(page 124)*. Then confirm the result with a dealer, explaining the size of the space you want heated, how often and for how long you expect to use the heater, and how many doors, windows and outside walls the room has.

Installation depends not only on type but on capacity and, to a certain extent, on the voltage for which the heater is designed. The smaller sizes of 120-volt units can, if necessary, be connected to an existing circuit provided it is not heavily used—some models simply plug in to a wall receptacle. But the more efficient 240-volt heaters and the larger 120-volt units require circuits of their own directly connected to the main service panel; such a separate circuit is recommended, in fact, for any heater. Connections at the service panel are best left to a professional unless you have had extensive experience in electrical work. However, the remainder of the installation and wiring—for either 120- or 240-volt heaters—involves only elementary techniques, and doing this much of the job yourself may save considerable expense.

The size of your heater will determine the size of the cable for the new circuit. Use size No. 12 for a heater—or a combination of heaters—using up to 3,000 watts, No. 10 for 3,000 to 4,000 watts and No. 8 for 4,000 to 6,000 watts.

All permanently installed heaters need either thermostats or switches. Many come with such controls already built in but others require separate controls, such as a timer switch *(page 76, Step 5)* for a ceiling heater, a remote thermostat *(page 19)* for a baseboard heater or an off-on switch for a kick-space heater.

A Unit to Recess into a Wall

1 Preparing the way. Find a wall space between two studs that is free of obstructions, and alongside one stud cut out a section of wallboard the size and shape of the heater housing shell. The bottom of the opening should be about 2 feet above the floor. Insert a drill in the fan housing opening and drill a ¾-inch hole through the sole plate and into the basement below.

If you are installing the heater in a room above other finished rooms, the easiest approach is to recess it in an interior wall that is directly above an interior wall in the room below. (Otherwise you may have to run cable from the heater along the inside of the wall through notches in the studs until you reach an inside-the-wall route to the basement.) Use an extension bit to drill through the sole plate and the top plate. At the floor below, cut out a section of wall near the floor and drill through the sole plate *(right)*. After installing the unit, patch this hole with new wallboard *(page 76, Step 6)*.

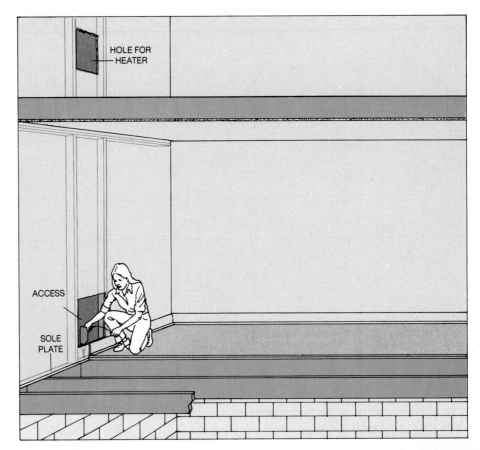

HOLE FOR HEATER

ACCESS

SOLE PLATE

ACCESS
HOLE

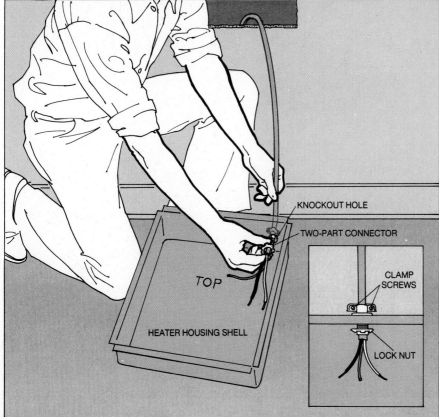

KNOCKOUT HOLE

TWO-PART CONNECTOR

TOP

HEATER HOUSING SHELL

CLAMP
SCREWS

LOCK NUT

2 **Fishing cable to the service panel.** Thread an electrician's fish tape through the hole in the wall and down into the basement, attach the cable to the tape and fish up about 3 feet of cable. In the basement, staple the cable to joists at 4-foot intervals along the route to the service panel. If necessary, drill ¾-inch holes through the center of the joists near the wall, run the cable through the holes, then drop the cable down to the service panel. Leave about 2 feet of cable at this end for the connection to the panel.

3 **Connecting the cable to the heater.** Strip about 8 inches of sheathing from the cable at the heater end, and about ½ inch of insulation from the ends of the two conductor wires. Clamp a two-part connector (*inset*) over the end of the sheathing, with the threaded end of the connector facing the stripped wires. Remove the twist-out from the knockout hole in the fan housing. Insert the wires and the threaded end of the connector into the knockout hole, screw the lock nut onto the connector from within the fan housing and tighten the lock nut with pliers.

4 **Connecting the heater.** After securing the housing shell to the stud at the side of the opening, prop up the fan-and-heater assembly so that the two screw terminals are at the top of the assembly and the terminals are within reach of the cable wires. If the heater runs on 240 volts, connect each insulated cable wire to a terminal (wrap black tape around the white wire to re-code it as a voltage-carrying wire). If the voltage is 120, connect the black cable wire to the terminal attached to the black heater wire and the white cable wire to the heater terminal with the white wire. Secure the bare ground wire around the green ground screw at the back of the housing shell. Screw the fan-and-heater assembly into the shell.

5 **Installing grille and frame.** If the grille and frame have been packed together, separate them. Screw the grille alone to the top and bottom of the fan-and-heater assembly, making sure that the assembly thermostat shaft protrudes through the center of the hole in the grille. Push the thermostat knob onto its shaft.

Place the heater frame face down on the floor and press the vinyl gaskets into place along the inner tracks at the edges of the frame. Hold the frame up to the heater with the gaskets facing the grille and push the frame into place.

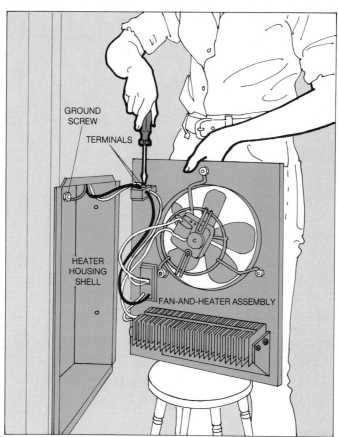

GROUND
SCREW

TERMINALS

HEATER
HOUSING
SHELL

FAN-AND-HEATER ASSEMBLY

THERMOSTAT
SHAFT

GRILLE

FRAME

VINYL GASKET

Putting a Timer-controlled Unit in the Ceiling

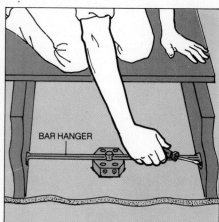

1 **Cutting a hole for the ceiling box.** Drill a ⅛-inch hole in the ceiling where you want to install your heater. If an unfinished attic is directly above, find the hole in the attic, center an octagonal electric-outlet box on the hole and draw its outline on the ceiling. Drill a starter hole to mark each corner. In the room below,

brace the surface of the ceiling with a block of wood and cut around the octagon.

If the space above is finished, install the box from below. Drill a small hole in the ceiling to find a free ceiling space, then cut a square hole that spans the distance between two joists.

2 **Installing the box.** Attach a ceiling box with one knockout removed to a bar hanger and screw the hanger to the joists, working from the attic or from below. Install the hanger with the edge of the ceiling box flush with the ceiling surface.

3 **Preparing to install a timer switch.** Probe to find a clear space in the wall for the timer switch. Trace the outline of a wall-outlet box over the hole and cut along the outline. If the space overhead is unfinished, put the switch on whichever wall is convenient, but if the space above is finished, locate the switch as shown here, so that you can run cable from it to the ceiling box between two joists.

Drill through sole plates and top plates as necessary (*pages 72-73, Steps 1-2*) and fish cable from the service panel to the switch hole.

4 **Fishing cable to the ceiling box.** If the space above the ceiling is an unfinished attic, drill a small hole in the ceiling above the switch location. Using this hole as a guide and working from the attic, drill a ¾-inch hole through the top plates directly above the switch hole, push a fish tape down, and have a helper in the room below snare this fish tape (*above*). Attach a cable to the first tape, fish the cable up through the hole in the wall and pull it over to the ceiling box.

If the space above is finished, fish the cable from below. Trace to the wall the two joists to

which you attached the ceiling box. Between them score a rectangle 2 inches long in the ceiling and an outline 4 inches long in the wall (*above*). Chisel out the wallboard within the rectangles. Fish cable from the wall opening through the two access holes and on to the ceiling box. Connect the cable to the box. Staple the cable to the top plates, near the top.

5 **Installing the switch.** Strip sheathing and insulation from the ends of the two cables (one leading to the service panel, the other to the ceiling fixture) and secure the cables with connectors in two knockout holes in a wall box (*page 73, Step 3*). Fasten the wall box into the wall opening. Using a wire cap, connect the two bare cable wires and a short length of wire the same size—a jumper wire—and attach the other end of the jumper to the ground screw in the back of the box.

For a 120-volt heater (*below, left*), connect a black wire from one cable to one switch terminal and the other black wire to the second terminal. Connect the two white wires from the cables with a wire cap. For a 240-volt heater (*below, right*), be sure to use a 240-volt switch with four terminals. Recode the white cable wires by wrapping their ends with black tape to indicate that they will be voltage-carrying wires. Connect the two insulated wires from one cable to the two lower terminals of the switch, the two wires of the other cable to the upper terminals of the switch. Screw the timer switch to the front of the wall box.

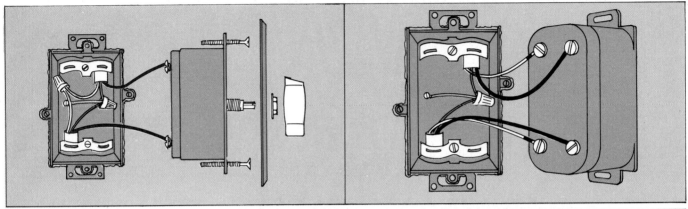

6 **Patching holes.** If you installed the ceiling box from above, proceed to Step 7. If you installed the ceiling box from below, you must patch the large hole in the ceiling before installing the heater. Cut two lengths of 1-by-2s a little longer than the exposed portion of the joists. Nail the 1-by-2s to the sides of the joists, flush with their edges. Cut a piece of wallboard to fit the opening and cut a hole in the patch to match the position of the ceiling box. Nail the wallboard patch to the 1-by-2s with wallboard nails. Patch the joints with joint cement, cover the cement with strips of perforated tape and cover the tape with three layers of joint cement; sand smooth after each layer dries. Fill smaller holes with spackling compound.

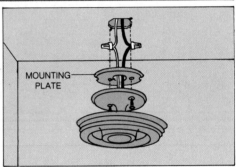

MOUNTING PLATE

7 **Installing the ceiling heater.** Slip the mounting plate over the heater assembly wires. Connect cable to heater wires with wire caps—black to black and white to white if the heater wires are color coded. Screw the mounting plate and heater to the ceiling box.

Ganging Units along a Baseboard

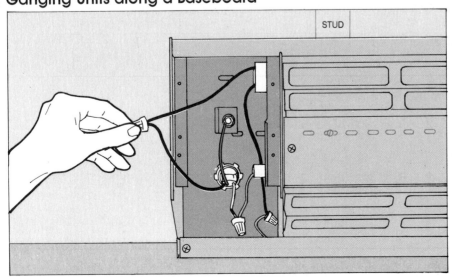

STUD

1 **Connecting the heaters.** Baseboard heaters are screwed to wall studs after removing the baseboard, and they are connected to the basement service panel as on pages 72-73, Steps 1 and 2. Inside a baseboard heater are two pairs of wires, one pair permanently joined at the factory with a plastic covering, the other joined by a wire cap. When connecting heaters, ignore the factory-joined wires but separate the other pair.

A 120-volt heater is wired to the cable by connecting the white cable wire to the white heater wire and the black cable wire to the black heater wire. Connect a 240-volt heater by attaching the black cable wire to either of the two heater wires. Then recode the white cable wire black and attach it to the other heater wire. For both 120- and 240-volt heaters connect the ground cable wire to the heater ground screw.

2 **Installing a second unit.** Remove the covers of both heaters. Remove the twist-outs from the knockout holes at the connecting ends of the two heaters. Screw the new heater to the wall studs and bolt the two units together through the small holes at the top of the heaters, using the nut and bolt supplied with the heater. Then push a bushing—also supplied with the heater—

through the matching knockout holes and secure the bushing with its lock nut *(below, left)*.

To wire two 120-volt heaters together, attach one end of a 10-inch-long jumper wire to the two black wires that you joined in the first heater. Pass the jumper wire through the bushing, into the second heater, and attach it to the black

wire of the second heater. Similarly, connect with a jumper wire the white wire on the first and second heaters through the bushing. To wire two 240-volt heaters together use a jumper wire to connect either of the two wires from the second heater to either of the two wire caps in the first heater. Connect the remaining wire on the first heater to the other wire cap *(below, right)*.

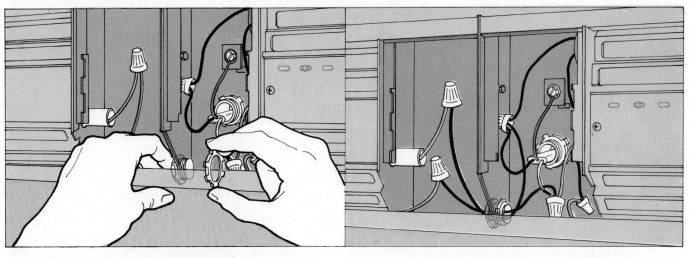

Tucking a Unit into a Kick Space

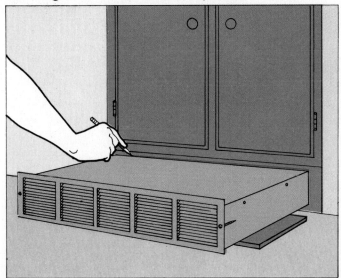

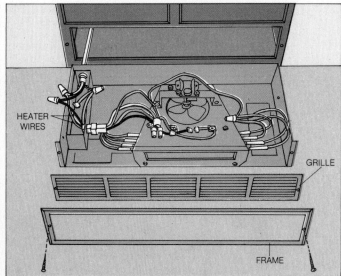

HEATER WIRES

GRILLE

FRAME

1 **Positioning the heater and switch.** Hold the back of the heater against the panel at the back of the kick-space and mark the outline of the heater along the panel. Saw out the outlined section. On a nearby wall find a location to which cable can be run conveniently and cut a hole for a wall box *(page 75, Step 3)*. Run cable from the basement to the wall-box opening *(pages 72-73, Steps 1-2)* and from the wall opening to the kick-space opening. Connect the cables in the wall opening to a side-clamped wall-outlet box and install the outlet box in the opening *(page 76, Step 5)*. Remove the top of the heater and connect the cable to the knockout hole in the back of the kick-space heater *(page 73, Step 3)*.

2 **Wiring the heater and the switch.** Use wire caps to connect one insulated cable wire to each of the two unconnected wires in the heater. Attach the bare cable wire to the green ground screw in the bottom of the heater. Screw the heater top back on and push the heater into the kick-space opening. Screw the heater frame and the front grille through the screw holes at the ends of the heater and to the wall behind. Install a standard on-off switch or a timer switch in the wall box as shown on page 76, Step 5.

The Fireplace: Decorative but Inefficient

Probably the most abused home heating system is the traditional fireplace, highly prized as a decoration and a mood-setter but notoriously inefficient as a heater—about two thirds of the heat in a fireplace is wasted up the chimney.

Still, a properly maintained and well-fueled fireplace can be a useful auxiliary to a central heating system, and can even reduce heating costs. Since each fireplace is usually sized to heat the room in which it is installed—a capability enhanced by the addition of convection grates (below, left) and heat-reflecting firebacks (right) —it is often possible to close system outlets in that room or to lower thermostat settings for the entire house if other rooms are not in use.

You can get more heat out of a fireplace by using the proper fuel. Coal gives more heat per pound than wood (chart, opposite), but most people prefer the crackle of a wood flame. Charcoal, although useful for outdoor barbecues, can be deadly indoors because it gives off odorless but lethal carbon monoxide. Hardwoods like hickory, oak or maple that have been air-dried for at least six months make the best wood fuels.

Many fireplaces work poorly for lack of adequate draft. This inability to pull in enough air results in slow-burning, smoky fires—especially in well-insulated homes, where it may be necessary to open a window for ventilation. Another cause may be a dirty chimney, which can be cleaned by lowering a gravel-filled bag down the flue and cleaning off the smoke shelf (page 80). Or there may be taller buildings or trees nearby, blocking the passage of draft-inducing breezes over the chimney top. The cure is to lengthen the chimney (page 81).

Poor draft can also be the result of incorrect fireplace design. The opening may be too large or too high—its height should never exceed its width—permitting smoke to roll out into the room. Check the dimensions against those recommended on fireplace charts, available from contractors and supply firms. If height is the cause of smoky operation, reduce the opening by building up the hearth with layers of firebrick (page 81).

Accessories that Deliver More Heat

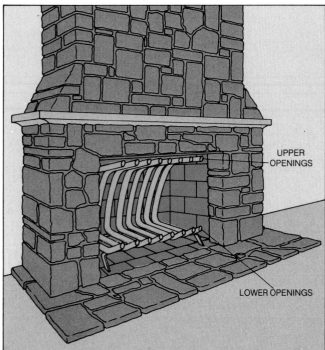

UPPER OPENINGS

LOWER OPENINGS

A convection grate. C-shaped hollow tubes draw in cool air at the bottom and send out fire-heated air at the top. The grate should be placed in the fireplace so that the upper openings are flush with the upper facing of the fireplace, with a gap of no more than 2 inches. If necessary, raise the grate by placing firebricks under the legs. Make sure that your fireplace screen does not block either the upper or lower openings when closed. You can improve the efficiency of the grate by attaching an electric blower unit to the lower openings.

BOLTING FLANGE

A metal reflector. A time-tested technique to get more heat from a fireplace is to attach a metal heat reflector, called a fireback, to the rear wall. You can use a sheet of aluminum flashing—or even heavyweight foil—but more attractive are ready-made firebacks of black iron about an inch thick, many of which are crafted in handsome designs. To install a fireback, place it against the back wall of the fireplace, with the bottom edge resting on the hearth, and attach it with lag bolts through the flanges into lead-bolt shields set in holes drilled into the masonry.

A Guide to Fireplace Fuels

What to burn. The upper section of the chart below lists the most widely used firewoods in the order of their heat value, measured in millions of BTUs to the cord—a stack of logs 4 by 4 by 8 feet. For comparison with other fuels, listed in the lower part of the chart, each kind of firewood is also rated in BTUs per pound.

Among coals, anthracite is the cleanest, hottest and longest burning, but is very difficult to ignite. Bituminous or soft coal is easier to ignite in fireplace grates, but may produce a sulfurous odor. Cannel coal, a type of bituminous coal that comes in large chunks, is a popular fireplace fuel because it lights, burns and crackles like wood—but with more heat.

At the top of the firewood list are high-density hardwoods, like hickory and oak, that burn slowly and provide an even heat. At the bottom are low-density softwoods, like pine and cedar, that burn hot and fast—producing more BTUs per pound but less per cord than some of the heavier hardwoods. Not all types of wood are available in all areas so your selection may be limited. Some woods provide unusual effects—fruit woods, such as black cherry, burn in a rainbow of colored flame while giving off an incense-like aroma, while the salt in driftwood collected along a seashore provides attractive dashes of blue and lavender to the flames.

Synthetic logs, made of compressed sawdust burn hotter and longer than wood logs but are considerably more expensive when used in quantity. The artificial logs made of newspapers, unless rolled very tightly, tend to burn quickly and unevenly and leave messy ashes.

Fuel	Heat value (Millions of BTUs per cord)	BTUs per pound (approx.)	Characteristics
Shagbark hickory	30.6	7,900	Burns completely without turning
White oak	30.6	7,900	Must be well seasoned; hard to split
Sugar maple	29.0	7,600	Burns fast
Beech	28.0	7,200	Colorful flame
Red oak	27.3	7,600	Needs two-year seasoning; must be turned frequently
White ash	25.0	7,400	Burns well without seasoning; splits easily
White elm	24.5	8,200	Burns poorly unless logs are very small; requires power saw for splitting
Red maple	24.0	7,500	Burns fast
Black cherry	23.5	8,400	Pleasant aroma
White birch	23.4	8,000	Burns completely
Douglas fir	21.4	8,600	Very smoky
Chestnut	20.2	8,100	Burns fiercely with frequent sparking; smoky
Spruce	17.5	8,300	Burns fast; smoky; sparks
Cedar	16.5	7,900	Pleasant aroma; smoky; sparks
White pine	15.8	7,500	Burns fast and hot without sparks; smoky
Sawdust logs	—	15,000	Burns evenly up to 3 hours; colorful flames
Newspaper logs	—	Variable	Burns well when tightly bound; smoky
Anthracite (hard coal)	—	15,000	Difficult to ignite; burns slowly with blue flame; sootless and smokeless
Bituminous (soft coal)	—	13,000	Burns fast with colorful flame; sooty and smoky
Cannel coal (soft)	—	10,500	Ignites quickly; sputters like a wood fire

Cleaning the Flue

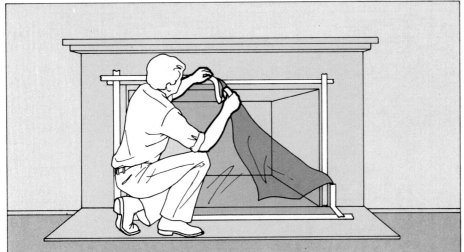

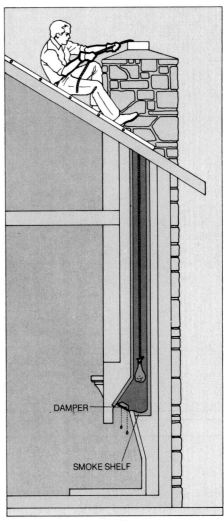

DAMPER

SMOKE SHELF

1 **Sealing the fireplace.** In preparation for cleaning the chimney, open the damper and separately tape at least two sheets of plastic over the fireplace opening, so that the longer second sheet completely covers the first *(above)*. Prepare a simple cleaning tool by weighting a small cloth bag, about half the size of a pillowcase, with two or three pounds of sand, gravel or chains. Around the bag neck, knot one end of a ½-inch rope long enough to reach from the chimney top to the fireplace throat.

2 **Cleaning the flue.** From the roof, preferably working from a ladder hooked over the roof ridge, remove the chimney storm cap, if there is one, and lower the bag into the flue. When slack rope tells you that the bag is resting on the smoke shelf at the bottom of the flue, draw it back up to the top. Repeat this two or three times, moving the rope position each time so that the bag brushes against a different portion of the flue.

Wait about an hour for soot and dust to settle before removing the fireplace covering. Start with a small opening at the bottom, so that the first updraft can sweep small particles up the chimney. Use a heavy-duty vacuum cleaner—available from a tool-rental firm—to remove the rest, or carefully sweep it up.

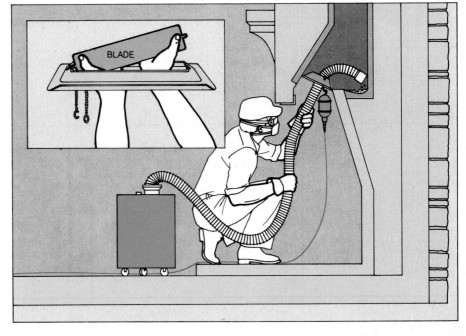

BLADE

3 **Cleaning the smoke shelf.** Don respirator, goggles and head covering for protection against soot. Remove furniture from the fireplace vicinity and cover nearby surfaces with dropcloths. Find a position from which you can reach upward into the fireplace throat. Then, using a work light for illumination, remove the damper blade from its pivot supports *(inset)*. On some dampers, you may have to first disconnect the rotary control at the front of the fireplace. If possible, insert the hose of a heavy-duty vacuum cleaner through the damper opening and bend it inward to clean the smoke shelf. If the hose won't reach the shelf, brush shelf debris through the damper opening, then vacuum it from the hearth. Replace the damper blade.

Extending the Flue

1 **Preparing the chimney.** Use a cold chisel and hammer to chip off the beveled cement cap that binds the flue liner to the bricks, working toward the outside so chips do not enter the flue.

2 **Laying the bricks.** Most chimneys have either two layers of brick or an outside layer of brick over an inside layer of concrete block. Begin with the outside layer. Use bricks that match the size and color of the old ones and lay each course from the corners inward. When the outside layer has reached the desired height—usually an extra foot or two—lay the inside layer of brick or cement block in the same manner (*below*).

FLUE LINER

CEMENT CAP

FLUE LINER EXTENSION

OUTER LAYER

INNER LAYER

3 **Topping off the chimney.** Use a power saw with an abrasive blade to cut a section of terra-cotta flue liner—available in 2-foot lengths—long enough to extend 4 to 8 inches above the new chimney top. Spread about an inch of mortar on the existing liner and place the flue extension on top of it. Trowel mortar from the chimney top into the spaces between the outer and inner layers of masonry, and between the inner layer and the flue liner. When the cracks are filled, trowel more mortar around the liner to make a mound about 4 inches high. Bevel the mound toward the chimney edges to form a protective cap. Wait at least two days for the mortar to set before lighting a fire.

Raising the Hearth

Reducing the fireplace opening. To determine the size of opening that gives the best draft, light a fire in the fireplace and slowly lower a large sheet of plywood across the face until no smoke escapes into the room. Measure the reduced opening. When the fire is out and the hearth cleaned, close the opening to the desired size by laying firebricks without mortar over the fireplace bottom, splitting them to fit snugly against the slanting side walls. Use up to three layers if necessary.

Installing Ready-made Fireplaces

You do not need a built-in masonry fireplace to enjoy the warmth and attractiveness of an indoor fire. Prefabricated metal fireplaces, manufactured in a wide range of sizes, shapes and colors, are designed for simple installation and are light enough—200 to 500 pounds—to be placed almost anywhere without requiring flooring supports.

There are two basic types of prefabricated fireplaces. One is an insulated metal firebox that can be placed against or inside a wall and faced with brick or other noncombustible material to resemble a conventional masonry fireplace. The other type is a freestanding fireplace with a decorative exterior, generally baked enamel, and factory-installed firebox, hearth and damper. Included in the freestanding category are hybrids like the Franklin stove, which doubles as a fireplace when its doors are opened.

When selecting a location for a freestanding fireplace, keep in mind the clearances required between fireplace and combustible surfaces—varying from a few inches to more than 2 feet—and the necessity of running a chimney past ceiling rafters or wall studs. The most efficient chimney is one that runs straight upward through ceilings, attic and roof. But it is often more convenient—and only slightly less efficient—to run the chimney pipe up an outside wall of the house to the roof, as shown in the eight-step sequence starting below.

This technique usually requires opening two holes through the side wall and roof overhang. But if the overhang or its attached gutter juts out 4 inches or less from the path of the pipe, you can use 15° elbows to clear the overhang.

The chimney itself is assembled from interlocking lengths of insulated pipe. Most fireplaces have enough decorative flue pipes to reach either an 8-foot ceiling or a nearby wall. Beyond these points, you must carefully plan the chimney run and assemble the necessary hardware: spacers to provide the minimum clearance of 2 inches wherever the chimney must pass through combustible material; collars to conceal pipe openings; Ts and elbows to change pipe direction; supports and bands to hold a chimney securely against the side of a house; flashing and storm caps to keep rain, insects and animals out. The chimney top should extend at least 3 feet above the roof.

A fireplace must rest on a noncombustible base at least ⅜ inch thick—brick or stone or a box of sand or gravel are common—to protect the surrounding floor from heat and sparks. This base will also keep unwary passersby from brushing against hot metal.

1 **Opening the wall.** Attach to the fireplace the section of decorative flue containing the damper and the damper control. Place an elbow on top of the flue as a positioning guide, the open end facing the wall. Measure the distance from the floor to the center of the elbow opening and mark at that height on the wall before removing the elbow. On frame construction, locate the adjoining studs on either side of the line and, if necessary, make a new mark midway between them. Cut a hole in the wall 2 inches larger than the flue, and make a similar hole in the exterior wall. Nail a wall spacer *(right)* at either end, centering the spacer holes in the round openings. If you use paneling, as shown here, cut an additional hole in the panel to be mounted over the openings.

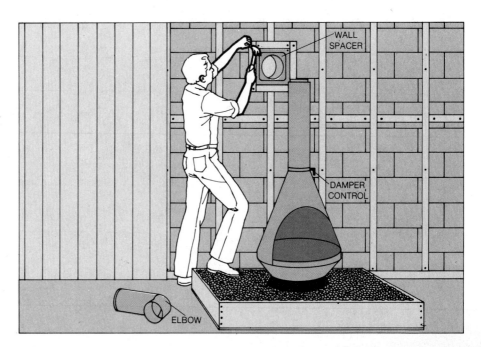

2 Connecting the flue. Replace the elbow and slide a trim collar onto the length of the flue that will connect the elbow and the wall opening. With the finished side of the trim collar facing you, push one end of the flue through the wall spacer until you are able to position the other end in front of the elbow opening. Connect the flue and the elbow (*right*). Slide the trim collar toward the wall until it completely conceals the opening. Fasten the collar to the wall.

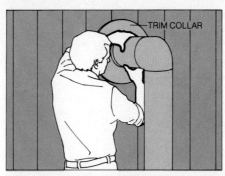
TRIM COLLAR

3 Mounting the wall support. Connect the chimney T to the plate of the wall support. Use the T as a positioning guide by inserting it into the wall and aligning it with the end of the flue pipe. It may be necessary to push the flue pipe back slightly—or to have a helper pull it back from inside—to allow the T to fit snugly, about 3 inches into the wall. Use a level to be sure the top of the T is perfectly horizontal, then have a helper mark the bolt positions through holes in the braces and bolting flanges (*right*). Screw the support to the wall with the diagonal braces above the plate, as here, or below it.

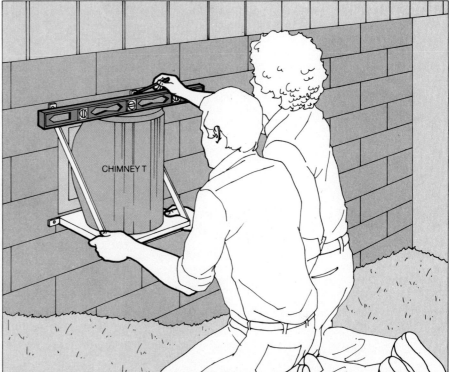
CHIMNEY T

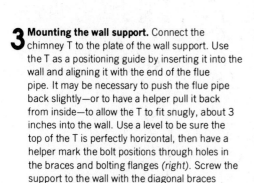

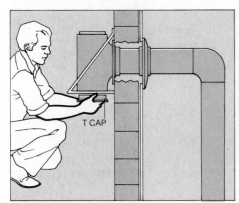

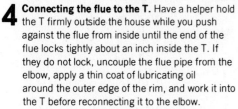
T CAP

4 Connecting the flue to the T. Have a helper hold the T firmly outside the house while you push against the flue from inside until the end of the flue locks tightly about an inch inside the T. If they do not lock, uncouple the flue pipe from the elbow, apply a thin coat of lubricating oil around the outer edge of the rim, and work it into the T before reconnecting it to the elbow.

When the flue has been firmly locked into the T, push in the T cap—used for inspection and cleaning—from beneath the support plate (*above*). Caulk around the T at the wall spacer.

WALL BAND
SCREW CLAMP

5 Running the chimney up the wall. Place the first length of insulated chimney pipe into the opening at the top of the T and twist it until it locks firmly. Place a level across the top of the pipe to be sure it is straight. Repeat the procedure, locking the bottom of each successive length of pipe to the top of the one below, until the chimney extends from the T to a point just below the roof overhang.

To anchor the chimney to the wall while retaining the required 2-inch clearance, wall bands should be attached at 8-foot intervals. Slide the wall band down the pipe to the proper position and tighten the screw clamp to fasten it to the pipe. Then nail the band to the side of the house (*left*).

6 **A passage to the roof.** If the center of the chimney pipe will be 4 inches or less from the edge of the roof overhang—check with a plumb line—use a pair of 15° elbows to carry the chimney around the overhang. Otherwise, mark the center line on the underside, or soffit, of the overhang and cut a hole in the same way as in an interior wall (page 82). If, despite earlier planning to route the chimney between rafters, the hole uncovers a lookout support, use an elbow to reroute the pipe. Use the plumb line inside the opening to mark the center for the second opening. Drill a guide hole upward through the roof. Cut the second hole from above, through shingles and wood. Nail a fire-stop spacer, similar to a wall spacer (page 82), to the soffit opening to provide for pipe clearance.

7 **Covering the roof hole.** Extend the chimney pipe through the holes in the soffit and roof until it reaches at least 2 feet above the roof peak. Spread roofing tar around the opening before sliding the flashing, which can be adjusted to the pitch of most roofs, down the pipe (below) until it lies flat against the roof and tar. Nail the flashing to the roof without removing shingles.

8 **Protecting the chimney.** Slip a storm collar over the chimney and slide it down to cover the opening where the chimney passes through the flashing. Tighten the collar's screw clamp. Then push a chimney cap onto the top of the chimney to keep rain, snow and birds out. Spread roofing tar around the top edge of the storm collar and the edges of the flashing.

Running a Chimney through the House

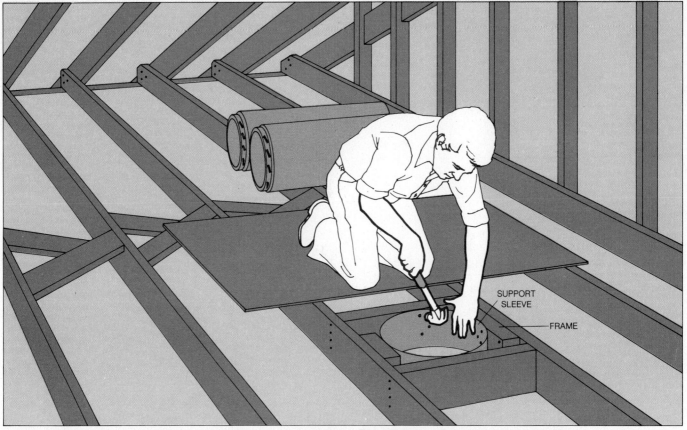

SUPPORT SLEEVE

FRAME

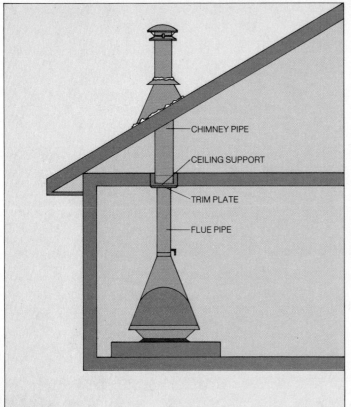

CHIMNEY PIPE

CEILING SUPPORT

TRIM PLATE

FLUE PIPE

1 Installing the ceiling support. Attach a length of decorative flue to the fireplace. Use a plumb line to align the center of the flue opening with a predetermined point on the ceiling midway between joists. If necessary, move the fireplace slightly to get the opening centered between joists. Cut a hole wide enough for the support sleeve in the ceiling as you would in a wall (*page 82*). From above, cut away flooring necessary to frame the opening with wood the same size as the joists. The frame should touch the circular opening on all four sides. From below, slide the support into the opening and nail the trim plate through the ceiling into the frame. From above, nail the sleeve to the frame (*above*).

2 Running pipe up. Insert a length of chimney pipe into the sleeve from above. Then connect it to the fireplace with flue pipe. Follow the steps used in an outside chimney run to open the roof and add chimney pipe to complete the job.

Where an inside chimney must pass through more than one ceiling, as for a fireplace located in a lower level of a multilevel home, only a spacer is needed in the second opening. But you should plan ahead to minimize the problem of running pipe through second-floor space. Route it through a closet, or run it through a corner where it can be boxed in like a duct (*page 58*).

Plugging the Sun into Existing Heating Systems

In a world running short of conventional fuels, sunshine is increasingly popular as an energy source. Solar systems like the one below heat water for household use. Others help to warm swimming pools *(page 91)* or houses. Many commercial units are available, and a good part of their cost is for installation—a fairly easy plumbing and electrical job.

Before you buy a solar water heater, make sure it will repay its cost in fuel savings. Dealers have this information, but you can calculate it yourself.

To do so you need your average annual ground-water temperature from your local water-supply authority. You also need the average annual amount of solar radiation for your locality in British thermal units (BTUs) per square foot. You can get that figure by writing to the National Heating and Cooling Center, P.O. Box 1607, Rockville, Maryland 20855, for "Solar Energy for Space Heating and Hot Water" (S.E. 101).

First multiply the average annual BTU figure by the area in square feet of the collectors you plan to install. Multiply the result by the efficiency of the collectors to get the annual BTUs of hot water the system will provide.

Next subtract the ground-water temperature figure from 140° (representing the normal temperature of hot water for general household use). Multiply the result by 125 (the pounds of hot water the average person uses daily). Multiply that result by the number of people in your family, then multiply again by 365. What you now have is the total BTUs needed to fill your annual hot-water needs. Compare that figure with the average annual BTUs of solar hot water that you calculated earlier. If the figure for solar BTUs comes within 40 per cent of your needs, a solar water heater could save you money.

Then make sure you have a proper site for the solar collectors—a yard, a flat roof or a south-facing pitched one that is not shaded during peak sun hours.

To catch the sun's rays efficiently, solar collector panels should be mounted at an angle, measured from the horizontal, that is within 15° of the latitude of the place where you live—a figure you can get from an atlas. For example, in an area at lat. 40°—Philadelphia, Pennsylvania, and Columbus, Ohio—collectors will function at any angle between 25° and 55°. The angle of any pitched roof will usually be within acceptable limits.

For all roof work, use a ladder made to clamp over the ridge of the roof or get a ridge-straddling attachment for your present ladder. Call a professional for any work on a slate roof or any roof pitched at more than a 45° angle.

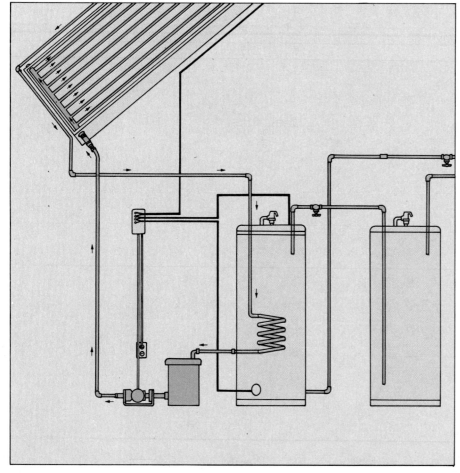

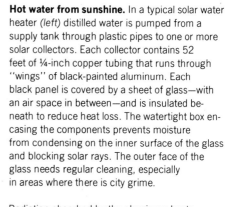

Hot water from sunshine. In a typical solar water heater *(left)* distilled water is pumped from a supply tank through plastic pipes to one or more solar collectors. Each collector contains 52 feet of ¼-inch copper tubing that runs through "wings" of black-painted aluminum. Each black panel is covered by a sheet of glass—with an air space in between—and is insulated beneath to reduce heat loss. The watertight box encasing the components prevents moisture from condensing on the inner surface of the glass and blocking solar rays. The outer face of the glass needs regular cleaning, especially in areas where there is city grime.

Radiation absorbed by the aluminum heats water flowing through the tubing. The sun-warmed water flows back to the house and through a copper coil inside a solar water tank that is linked with the existing household water heater. Water in the coil transfers heat to water in this solar tank, then is pumped through the system again. Warm water is drawn from the solar tank into the house hot-water tank, where it is heated further if necessary. (During periods of little or no solar heat, the existing gas or electric unit supplies most or all of the heat.) Water from the main that supplies the house refills the solar tank as it empties.

Thermal sensors, one attached to a collector panel and two attached to the solar tank and wired to the pump, help to keep the tank's temperature constant by controlling the flow through the closed part of the system.

Putting Collectors on Pitched Roofs

1 **Locating the collectors.** Determine where on the roof you want to put the collectors. They should be as close as possible to the water heater. From inside the attic, find a rafter near where one outside edge of the line of collectors will fall. Drive a tenpenny nail through the roof next to the rafter and 2 feet above the attic joists. Note the spacing of the rafters—usually 16 or 24 inches from center to center.

On the roof, find the protruding nail and measure over from it ¾ inch to find the center of the rafter. Mark its position with chalk and similarly mark the positions of rafters along a horizontal line the length of the row of collectors. Make parallel marks above the first at a distance that is the height of the collectors plus 1 foot. Remove the nail and patch the hole.

2 **Preparing collector supports.** Place the collectors on the ground side-by-side, as they will be when installed. Solder copper caps over the mouths of all other manifolds except the inlet and outlet connections and those that you will connect to each other (*page 88, Step 6*). Cut two lengths of 1-inch aluminum angle as long as the row of collectors. At 1-foot intervals, drill holes through one side of each strip of angle and corresponding holes through the collector frames. Open the trap door in the collector that now has a reducer at the upper outside corner. With an awl, punch a hole through the aluminum collector plate near the manifold outlet and attach an electronic thermal sensor to the underside with a sheet-metal screw (*inset*).

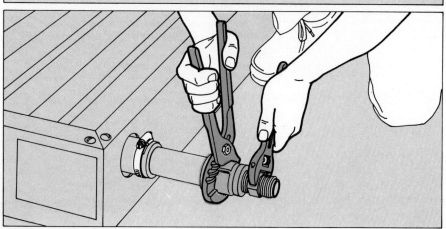

3 **Installing reducers.** Slip a 4-inch length of rubber hose over the manifold at the bottom outside corner of the collector at one end of the row. Insert a 4-inch length of 1½-inch copper pipe into the hose. Clamp the hose to the manifold and the pipe with hose clamps. Solder a copper reducer to the pipe. Screw a plastic union into the reducer with two layers of plastic joint tape wound tightly around the threads at the reducer end. Gently tighten the connection with two wrenches. Water will enter at this connection when the collectors are installed. Similarly attach a hose, a pipe and a reducer to the top outside corner of the collector at the other end of the row to make a water outlet. If the system is self-draining, install the drain device at this point.

4 **Preparing the roof.** The outlet end of the collectors must be tilted downward at a slant of about 1 inch in 20 feet. Attach one end of a chalkline where the bottom corner of the inlet end will be. Attach the other end of the line 20 feet away and an inch lower, and snap the line. Place a strip of aluminum angle along the line and drill through the angle and roof into the rafters. Fill the holes with roofing compound and attach the angle with 2-inch flathead wood screws. Cover the screwheads with roofing compound.

5 **Installing the collectors.** Tie a rope securely around a collector, leaving two long rope ends. Using two roof-mounted ladders as secure footing for yourself and a helper, haul the collector up to the roof on a third ladder as shown at right. Set the collector in place on the aluminum angle, line up the holes in the collector and the angle and fasten the collector to the angle with sheet-metal screws. Similarly, haul up a second collector and place it next to the first one without attaching it.

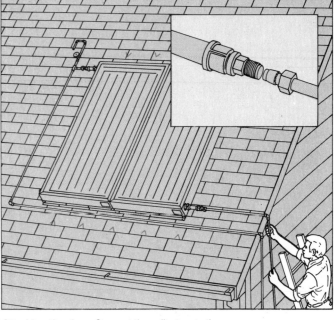

6 **Connecting the collectors.** Open the trap doors at the tops and bottoms of the casings to expose the manifolds, the large copper tubes at the top and bottom of each collector. Connect the two top manifolds with a 2-inch piece of 1½-inch rubber hose held at each end with hose clamps. The ends of the manifolds should be about ⅛ inch apart to allow for expansion. Similarly connect the two bottom manifolds. Screw the bottom of the second collector to the aluminum angle. Screw the other aluminum angle to the rafters along the tops of the collectors and bolt the angle to the collectors.

7 **Attaching the pipes.** Connect the collectors to the system with ½-inch polybutylene pipe. Connect piping to the inlet and outlet fittings. Insert a plastic stiffener in a length of pipe and slip a plastic compression nut over the end. Fit a plastic ferrule over the end of the pipe (*inset*), slip the pipe inside the plastic fitting and tighten the nut. Using elbow joints as necessary, extend piping from the outlet, down the side of the collector, along the bottom of the collectors and down the side of the house as shown. Extend piping from the inlet down the side of the house. The pipe from the outlet should slope downward to the roof edge.

8 **Hooking up the solar system.** Drill two ¾-inch holes through the house foundation at least 2 feet above the ground, extend pipes through the holes and seal with butyl rubber caulking. Extend the outlet pipe to the solar water tank, strapping it to joists as necessary, and join it to the inlet atop the solar tank with a reducer. Join the solar tank outlet to the supply tank, the supply tank to the pump and the pump to the inlet pipe.

9 **Hooking in the existing system.** Turn off the water in the pipe carrying cold water to the existing heater. Remove the elbow that connects this pipe with the existing heater, leaving 6 inches of pipe protruding from the top of the existing heater. Extend a line of ¾-inch pipe to the solar heater, bringing it down to the inlet with an elbow. Similarly, run the pipe to the heater inlet with an elbow and join it to the tank with a union.

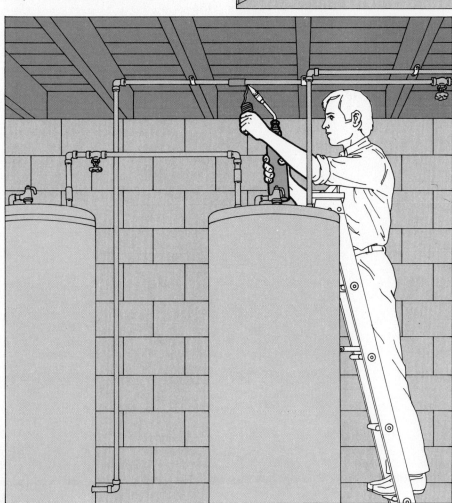

Making the Final Connections

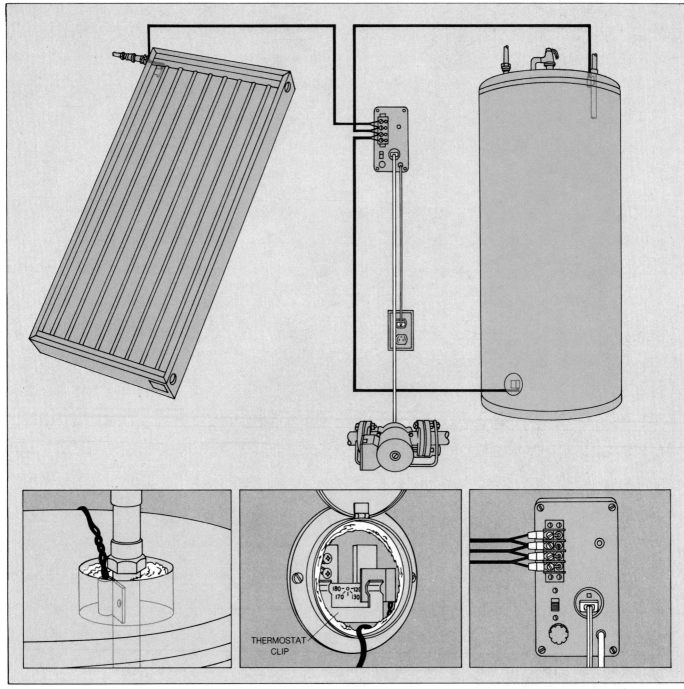

THERMOSTAT CLIP

Installing the electronic control. One of the three temperature sensors necessary to regulate the water-heating system is screwed to a collector (*page 87, Step 2*). To install the upper limit sensor, first remove insulation from around the outlet pipe at the top of the solar tank. Disregarding the screw hole, place the sensor against the pipe and pack the insulation around the sensor to hold it in contact with the pipe.

Remove the thermostat cover plate at the bottom of the solar tank. Slip the sensor behind the

thermostat clip to hold it in contact with the tank. Replace the cap.

Mount the panel on a wall near the pump. Run the pairs of wires from the three sensors to the control panel and then attach them to the terminals with spade connectors. The connectors, as well as the inexpensive tool that is used to crimp them to the wires, are available at electric-supply stores. In this case, two of the connectors hold a single wire each and two connectors hold two wires from different sensors.

Using a wire cap, connect the control-panel wires marked CONTROLLED OUTPUT to the leads of the pump and plug the power cord from the control panel into a grounded 120-volt receptacle, independent of other electrical outlets. Fill the supply tank with distilled water, and turn on the water from the main; correct any leaks in the system. Finally, cover all pipes with foam-rubber pipe insulation. Before insulating the outlet pipe from the collector, tape the wires leading from the sensor on the collector to the pipe so that they will be covered by the insulation.

Installing a Solar Pool Heater

You can easily add a solar system as the sole heat source for a swimming pool or as an auxiliary heater to economize on conventional fuel. The plumbing required is the same for both installations and is less complex than for the water heater described on pages 87-90, since a swimming pool already has a pump to circulate and filter the water. All you have to do is install a valve and Ts to divert the water through the collectors on its way back to the pool. The solar collectors for warming a swimming pool are lighter and simpler than those needed to heat household water since pools are heated only to about 80° instead of the 140° of household water. The panels are not encased or covered with glass but consist merely of copper piping covered by aluminum collector plates.

The number of collectors needed depends on the size of the pool and its geographic location. In Zones 1, 2 and 3 of the map on page 124 the collectors should cover an area that is one third the size of the pool; in Zone 4, one half the pool size; and in Zones 5 and 6, three fourths the pool size.

Installing such an expanse of collectors in a yard is usually impractical. When no suitable roof is available, many homeowners have built simple cabanas with roof areas large enough to hold the relatively lightweight collectors.

1 Anchoring collectors to a roof. Locate and mark rafters as in Step 1, page 87, and snap a slanted chalkline along the top marks. Along the line drill a hole in each rafter for a 2-inch lag bolt. Fill the holes with roofing compound. Seal manifold openings that will not be connected, as in Step 3, page 87. Pull the first collector into place, hook one-hole straps over the top manifold and attach the straps to the rafters with lag bolts. Bring up the additional collectors and hook them together as in Step 6, page 88, then bolt them to the roof. Connect the inlet and outlet manifolds to the polybutylene pipe as in Step 7, page 88 and run the pipe to the existing pump and filter system.

2 Connecting the collectors. Attach a T to each side of a gate valve with short lengths of pipe. Cut a section of the pipe from the return line to the pool—at a point before the line enters an existing heater, if there is one—and replace the section with the valve-and-T assembly. Join the collector inlet to the T nearest the pump and the outlet to the T nearest the pool. By opening and closing the valve, you can regulate the flow of water to the collectors—and pool temperature. Or you can install an automatic control unit similar to the one for a hot-water system shown on page 90.

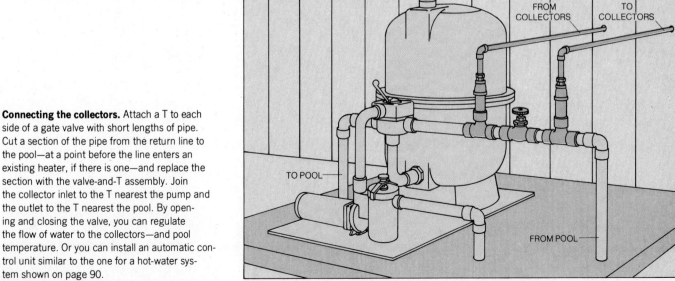

FROM COLLECTORS

TO COLLECTORS

TO POOL

FROM POOL

FAN OFF COOL

HI HI

M
E
D

M
E
D

LO LO
SLUMBER SPEED

4

Seven Ways to Cool a House

The obvious method for keeping your house comfortable in summer, whether you live in Quebec or Death Valley, is to install the refrigeration machinery and ducts of a central air-conditioning system. It is the most effective method. It is also fairly simple—ready-to-install kits can add an air-conditioning unit to a warm-air furnace *(pages 108-117)* and a small attic unit can be installed in many houses that do not have existing ductwork. But central air conditioning is only one of at least seven devices that can mechanically cool your home.

Room air conditioners, though noisier than a central system, are cheaper and easier to install. A single unit can cool some small homes and you can keep a larger house comfortable by combining one or two strategically placed units with portable fans *(pages 106-107)* to circulate cool air throughout the house.

Just removing hot air helps too. Small roof and gable fans *(pages 94-97)* lop up to 30 per cent off the operating costs of a central system by exhausting the hot air that builds up in an attic and leaks heat to rooms below. A larger attic fan *(pages 98-99)* that sucks cool night air into the house through open windows can help even more and can often cool a whole house unaided.

Evaporative coolers *(page 101)* are cheapest of all. The water vapor they add to the air they cool is a boon in hot, dry climates but limits their use in more humid regions.

But installing machinery is only half the battle. You can keep summer heat and humidity from getting inside the house by sealing it carefully with weather stripping, insulation and storm doors. Do whatever you can to block the heat of direct sunlight from entering. Using light-colored roofing—or painting the existing roofing white—reflects heat away. Old-fashioned shutters or overhangs called eyebrows, added to shade windows from the sun, stop solar radiation before it passes through the window glass and is trapped inside by the "greenhouse effect." And remember that nature is an ally. Landscaping *(pages 48-48G)* can help a great deal to make your house comfortable in summer—trees and vines make excellent sunshades, and plantings can be arranged to make the most of cooling breezes.

Some of the most effective cooling measures involve only changes in household routine. Open windows after sunset to allow naturally cooled air to circulate; close shutters and windows and draw blinds to keep daytime heat out. With the house sealed against outside heat, control internal sources of heat and moisture. Turn off unnecessary lights and use irons and hair driers sparingly. Make sure clothes driers are vented outdoors. To prevent heat and humidity from flowing to other rooms, close kitchen and bathroom doors while cooking or showering; then use vent fans to exhaust the heat.

Cooling a House without Using Refrigeration

On a sweltering summer day, a house may seem uncomfortably warm even though the air conditioner works constantly—and expensively—to cool it. In the same weather a house without an air conditioner will have a lower electric bill but may be unbearable to live in. For both houses, the best solution to the problems of cooling and its costs may well be a system of attic fans and vents.

In the air-conditioned house, a roof fan could suck hot air out of the attic, or a fan in an attic gable could draw cool outside air across the attic from a vent in the opposite gable. In a house without air conditioning, a fan mounted in an attic floor (pages 98-99) will pull cool outdoor air into the entire house to cool both the living quarters and the attic.

Attic ventilation in an air-conditioned house can be almost as important as the air conditioner itself. Air trapped in an unvented attic may get as hot as 150°. This heat seeps to the living quarters below, making the air conditioner work harder, but to less effect. An attic temperature of 135° may force the air conditioner below to run continuously to keep the living quarters at 78°. Aided by a roof fan that cuts the attic temperature to 95°, the same air conditioner may run only intermittently—and the bill for its power may drop by as much as 30 per cent.

The roof fan below is housed in an assembly mounted over a hole in the roof. The assembly includes more than a motor and blades: it has flashing to prevent water from leaking through the hole, a bubble-like cover that protects the fan motor from rain and snow, and a screen to keep out birds and large insects. A thermostat turns the fan on and off at preset temperatures that can be adjusted to meet your needs.

When installing a roof fan, be sure to follow the basic rules for safety in high places (page 86). The installation itself is both safe and easy in an asphalt shingle roof, but may split or break wood or slate shingles and is impractical in a metal roof. As an alternative you can install a gable fan, relatively slow in cooling an attic but safe to install in any house. If the attic gable has no vent, the fan is mounted over a special vent with louvers that open by air pressure, then close when the fan turns off (page 97). If you already have gable vents you can install the fan over one of these vents (page 97).

The capacity of the roof or gable fan you install, measured in cubic feet per minute (cfm) at "static air pressure," depends on the size of your attic and the color of your roof (a dark roof absorbs more heat than a light one, and calls for a more powerful fan). To determine the minimum cfm, multiply your attic floor area by 0.7, then add 15 per cent for a dark roof. For example, an attic measuring 2,000 square feet requires either a 1,400 cfm fan under a light roof, or a 1,610 cfm fan under a dark. (If your fan is rated at "free air delivery" rather than static air pressure, discount its cfm rating by 25 per cent to arrive at the right figure.)

Whatever the size of the fan you use, your attic must be equipped with vents—ideally, soffit vents at regular intervals under the eaves for a roof fan, both gable and soffit vents for a gable fan. For every 150 square feet of attic-floor area, install at least 1 square foot of "net vent area"—that is, of a vent that offers no more resistance to the flow of air than ½-inch screening. Vents with metal louvers call for 1½ times the net vent area; with wood louvers, twice as much.

Fitting a Fan into a Roof

1 **Positioning the fan.** Assemble the fan and carry it up to that part of the house roof that faces away from the street, and near one of the gables. Using a piece of wood as a guide, set the assembly at a point where the top of the fan cover is level with the roof ridge. Measure the distance from the ridge to the center of the fan.

In the attic, locate the two central rafters of the back roof—midway between gables—and, measuring down from the roof peak, locate the point halfway between them that corresponds to the desired fan position. Drive a long nail up through the roof at this point.

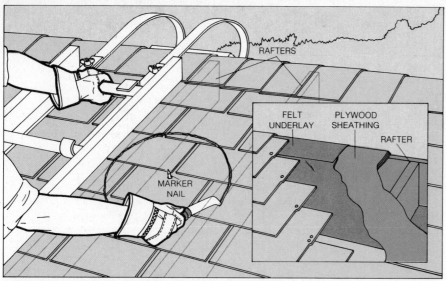

2 **Removing shingles and underlay.** On the outside of the roof locate the marker nail and, using it as a center, mark a circle on the asphalt shingles about 4 inches wider than the hole specified for the fan in the manufacturer's instructions. With a linoleum knife, cut along the outline of the circle until you reach the board or plywood sheathing underneath. Remove the nails holding the shingles and underlay in place, and discard the free shingles and underlay.

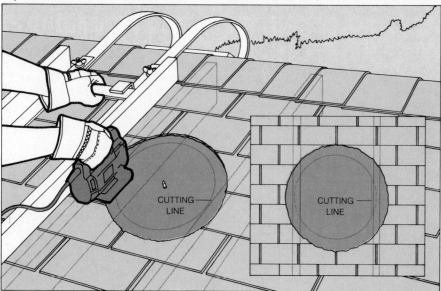

3 **Cutting through the sheathing.** Using a piece of string tied to the marker nail at one end and a pencil at the other, draw the outline of the hole for the fan on the wood sheathing, then cut the hole out with a saber or a keyhole saw. Caution: the hole specified by the manufacturer may be just wider than the distance between the two adjoining rafters; if so, do not cut through the rafters but saw along their inner edges (*inset*).

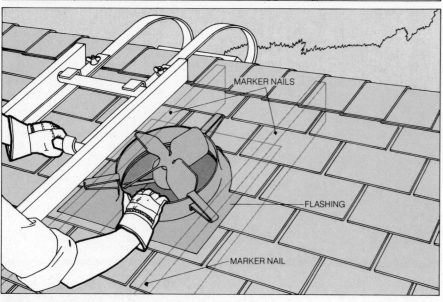

4 **Installing the fan housing.** From the outside, hammer four nails partway into the roof to indicate the centers of the adjoining rafters just above the area to be covered by the fan flashing, and about 6 inches below. Apply roofing cement liberally to the sheathing you have exposed and to the underside of the fan flashing. Remove the fan cover so that you can see through the hole in the fan housing, and slip the flashing up under the shingles until the housing hole lies directly over the roof hole; if necessary, use a pry bar to remove any shingle nails that bar the way. Using the marker nails as guides, drive galvanized steel nails through the flashing and into the rafters below at 3-inch intervals. Remove the marker nails.

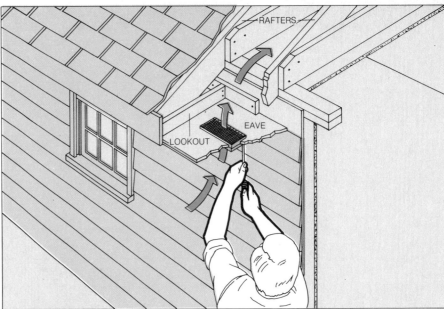

5 **Installing the screen and cover.** Slide the cylindrical screen into the fan opening and secure it to the hooks on the sides of the opening. Set the cover over the opening and bolt it to the flanges at the rim of the housing.

6 **Installing soffit vents.** Calculate the vent area that you will need for your attic (*page 94*) and the number of soffit vents—most vents are 8 by 16 inches or smaller—required to make up this area. Install the vents at regular intervals under the eaves on both sides of the house. To install a vent, hold it against the underside of the eave midway between the outside wall and the edge of the eave, and midway between two of the lookout beams that support the soffit (you can locate a lookout by the exposed nails that hold the soffit to it). Outline the screened area of the vent on the soffit and cut out the outlined area with a saber or keyhole saw. Insert the vent into the opening from below and screw it to the soffit through its flange holes.

7 **Wiring the fan.** In the attic, fasten the fan thermostat to a rafter, with its control dial readily accessible and the temperature-sensing element on its back exposed to the air. To power the fan you must use a 120-volt junction box in the attic or a receptacle on an inside wall in a room below. For the latter location, drill a ¾-inch hole through the top plates directly above the receptacle. Turn off the power to the receptacle and remove it from the outlet box. Fish cable from the thermostat to the receptacle through the hole in the top plates; clamp one end of the cable to the thermostat and the other end to the receptacle outlet box (*page 73, Step 3*).

Using wire caps, connect the black wire of the cable to the black wire of the thermostat, the white cable wire to the white thermostat wire, and connect the bare wire to the ground screw. To connect the other end of the cable to an end-of-the-run receptacle, attach the black and white cable wires to the free terminals, and join the bare wire to the existing bare wire with a wire cap, run a jumper wire from this wire cap to the receptacle ground screw. To connect the cable to a middle-of-the-run receptacle (*inset*), remove a black wire from its terminal and join it to both the new black wire (*dash lines*) and a black jumper wire (*dash lines*) with a wire cap; attach the jumper wire to the terminal. Connect the new white wire (*dash lines*) in the same way. Attach the new ground wire (*dash lines*) to the other bare wires and to the jumper wire.

Opening a Gable End for an Attic Fan

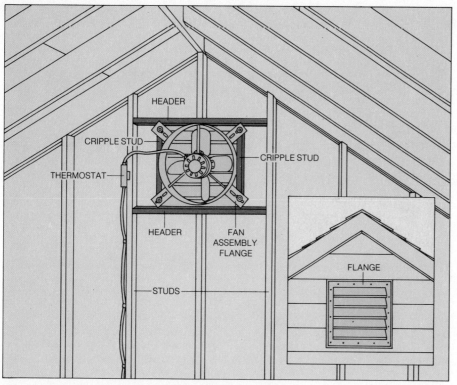

Cutting a hole for the vent. Inside the attic, outline the hole for the fan vent as high as possible at the center of the gable. In a house with clapboard siding, cut out the outlined area, including the central stud, then cut an additional 1½ inches from the stud above and below the opening. Cut two 2-by-4 headers to fit horizontally between the flanking studs and nail them to the cut stud above and below the opening and to the flanking studs. Cut two cripple studs to fit between the headers and nail the cripples to the headers along the edges of the opening.

Hold the fan assembly in the framed opening with the oil holes of the motor pointing upward and nail the four assembly flanges to the cripple studs. Screw the thermostat to a stud and wire it to a receptacle *(left, Step 7)*. Outside the attic, insert the fan vent in the opening and nail it to the siding through the flange *(inset)*; caulk the edge of the vent flange with roofing cement. Install a standard vent, framed and caulked in the same way in the opposite gable.

In a brick wall, drill holes from inside the attic at the corners of the vent area. Outside the attic, connect the holes with a brick chisel and a hammer; remove the bricks within the outline. Secure the vents with masonry nails.

Mounting Fans on Existing Gable Vents

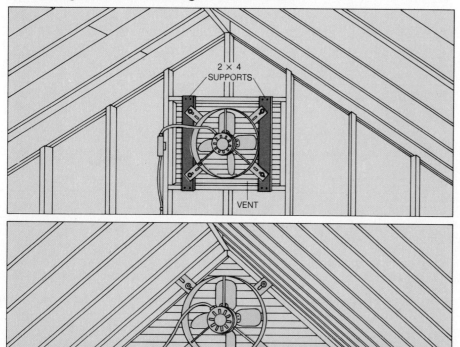

Over a rectangular vent. Set the fan at the center of the vent and mark the width of the fan cylinder on the headers above and below. Cut two 2-by-4 supports to fit the vertical distance between the marks, and nail them in place with their inside edges flush to the marks. Screw the ventilator fan to these supports and the thermostat to a convenient stud; connect the thermostat to a receptacle *(left, Step 7)*. If the gable vent at the other end of the attic is too small to serve the fan *(page 94)*, add soffit vents *(left, Step 6)*.

Over a triangular vent. If the fan flanges extend to the wood vent frame, screw the assembly directly to the frame. Otherwise, install vertical 2-by-4 supports as you would for a square vent, and attach the assembly to these supports.

A King-sized Fan to Cool a House

Almost anywhere in the U.S. and Canada, a home without air conditioning can be effectively cooled by a single powerful fan, mounted on the floor joists of an attic and controlled by a switch downstairs. The fan pulls cool, fresh air into the house through open windows and drives hot attic air out of the house through gable and soffit vents. Less expensive than central air conditioning, the fan is also much cheaper to run.

The fan should be centrally located, or near a stairwell if there is one. A louvered shutter in the ceiling below automatically closes when the fan is off, sealing off hot summer air. The fan assembly itself houses blades from 24 to 42 inches long and rests on a gasket to reduce vibration.

The size of the assembly determines the construction of the wood frame that supports it. When the spacing of the joists matches the size of the fan blades, the frame consists of a square formed by two joists and two headers. When the assembly is wider than the space between two joists, the frame is more complex (right and on the following pages).

For effective cooling, a fan should make one complete change of air per minute in a house in the South, or one change every two minutes in the North and Canada. To determine how much air must be moved in your house, multiply total floor area by ceiling height and subtract 10 per cent for closets and other unused space. This figure gives you the net air volume of the house and is the minimum air-moving capacity, measured in cubic feet per minute (cfm), required of a fan in the South; a fan in the North needs half that capacity. For example, a house with a net air volume of 12,000 cubic feet requires a 12,000 cfm fan in the South, a 6,000 cfm fan in the North.

The fan you buy should have two safety devices: an automatic shutoff for an overheated fan, and a control that shuts the fan down and closes the shutter's louvers to seal off the attic during a fire. Ask the dealer for the correct attic vent area for your fan and calculate the net area for metal or wood-louvered vents (page 94). Methods for installing vents are shown on pages 96 and 97.

Installing a Fan in the Attic Floor

1 **Positioning the fan.** Place the shutter upside down on a sheet of cardboard, trace the outline of the area that will fit into the ceiling and, using a utility knife, cut this piece out as a template. Drive a nail through the ceiling at the location you have chosen for the fan. In the attic, set the template over the nail with one edge flush to the inside edge of a joist. Mark the corners of the template on the surface below, and drill small holes at these points.

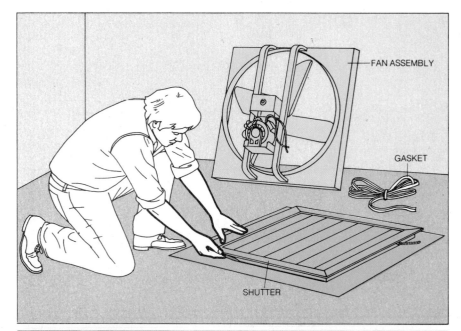

FAN ASSEMBLY

GASKET

SHUTTER

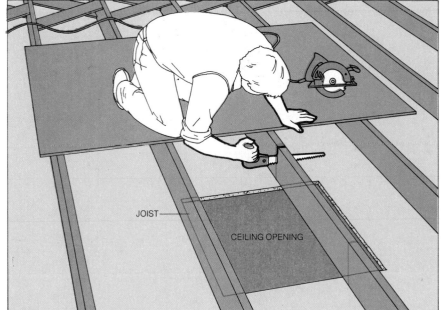

JOIST

CEILING OPENING

2 **Cutting the opening.** In the room below, join the four holes with straight lines to form a square outline and, wearing goggles and gloves, cut along the lines with a keyhole or saber saw. If you meet a joist, do not cut through or under; skip the joist area and resume cutting along the outline beyond the joist. Score the uncut segments of the outline with a utility knife, break up the ceiling board under the joist with a hammer and tear it away with your hands. Working in the attic, cut out the exposed joist 1½ inches outside the edges of the opening. Cut through most of the joist with a circular or saber saw, then, for the last cuts, use a keyhole saw—with the handle reversed, to avoid damaging the ceiling below.

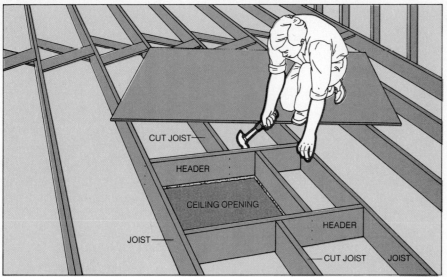

3 **Framing the opening.** Measure the distance between the uncut joists flanking the opening and cut two headers to this length, using wood the same size as the joists. Set the headers at the edges of the opening, and nail them to the sides of the uncut joists and the ends of the cut ones. To complete the frame, cut a third length of wood to fit between the headers, set it flush to the unframed edge of the opening and nail it to the headers.

CUT JOIST

HEADER

CEILING OPENING

HEADER

JOIST

CUT JOIST   JOIST

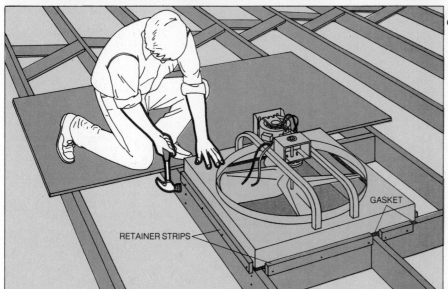

4 **Installing the fan assembly.** Set the felt or rubber gasket on the wood frame (in some models this gasket is attached to the fan assembly at the factory) and then lower the fan onto the cushioned frame. Nail 1-by-2 retainer strips to the sides of the joists and headers around the fan assembly, with about an inch of each strip projecting above the frame.

GASKET

RETAINER STRIPS

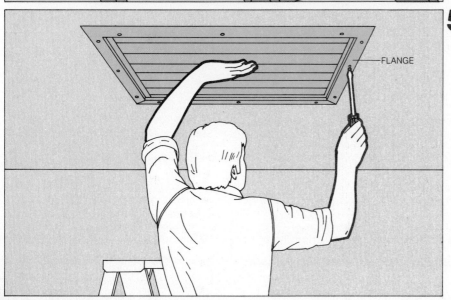

5 **Completing the job.** In the room below, adjust the spring at the side of the shutter to close the louvers gently when you hold them open and release them. Working from a stepladder, lift the vent into the ceiling opening and mark the positions of the screw holes in the shutter flange. Drill pilot holes at these points and screw the shutter through the ceiling to the frame above.

In the attic, connect the black and white fan motor wires to a cable in an outlet box (*page 73, Step 3*) fastened to a nearby joist or header; from the box, run this cable to the basement service panel for a new 20-amp, 120-volt circuit connection (*pages 72-73, Steps 1-2*). Install an on-off or a timer switch (the installation procedure is identical) on a convenient wall in the living quarters (*page 76, Step 5*).

FLANGE

Cool Air from a Furnace

Most forced-air furnaces have a switch that lets you turn on the blower without turning on the burner *(page 9, box)*. The blower then keeps air circulating back and forth through the house to provide relief on hot days. You can also use the fan switch to bring into the living quarters cool air that accumulates in a basement (it will be cooler than the house air but no less humid).

To accomplish this, you can simply remove the access panel covering the front of the blower. But to avoid pulling in dust from the basement, a better method is to modify the access panel. Cut a hole in it and install a filter over the hole *(below)*. At the beginning of the heating season, replace the filter with a plywood panel to close the opening so that the heating system will operate normally, recirculating warmed air.

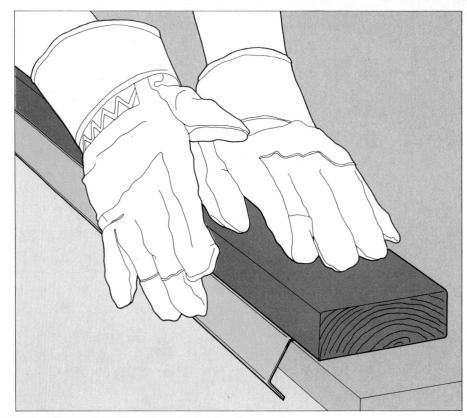

1 **Making a filter holder.** In attaching a new filter to a furnace blower for summer use, you will want to use metal channels, preferably J shaped in cross section for easy installation. If your dealer has no channel in the right size, you can make some yourself by bending thin (26 gauge or so) sheet metal, as shown above. Place a strip 2¾ inches wide with its edge extending ½ inch over the edge of a bench or table. Hold or clamp a board over the metal and bend it to a 90° angle. Make a second 90° bend 1¼ inches in from the first.

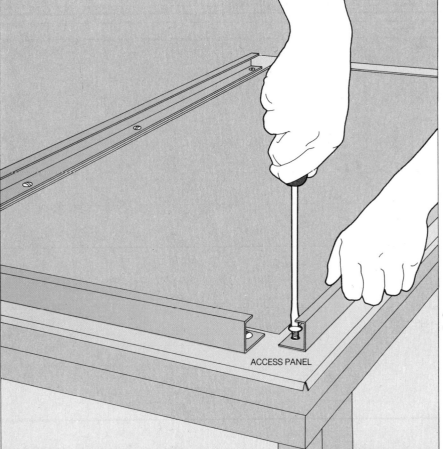

ACCESS PANEL

2 **Attaching the channels.** Remove the access panel and cut a hole in it an inch or so smaller all around than the filter you will install. Lay the panel on a bench or table and fasten channels along the sides and bottom of the opening with sheet-metal screws. Position the channels so that the filter will slide in easily between the side channels and rest on the bottom channel. Apply self-sticking insulating foam strips around the edge of the opening to cover the screwheads and provide a good seal for the filter.

Thrifty Cooling for Dry Climates

The ancient Egyptians, aware that as water evaporates it draws heat from the surrounding air, hung soaking wet reed mats in doorways to cool entering breezes. The modern evaporative cooler works on the same principle, using fiber pads instead of a mat and a blower to supply the breeze. Water enters a reservoir in the unit through a float-operated valve. When the reservoir is full, the float closes the valve to stop the flow. A pump forces the water through tubes onto the pads. Most of the water evaporates into the air that is drawn through the pads by the blower; excess water drips from the pads back into the reservoir.

Evaporative coolers work only in dry climates like the American Southwest—since they add moisture to the air they would make a sticky day worse in humid areas. But in arid regions the humidity they add to the air is welcome.

The coolers have other advantages to offer, too. Because a cooler does not have a compressor, it is less expensive both to buy and to operate than an air conditioner—it uses about a quarter as much electricity.

The air pumped into the house by the cooler must have an exit—usually opened windows—to keep the cool air flowing. The cooler's pads should be changed once a year—or twice a year in areas where hard water clogs them with mineral deposits. Using artificially softened water does not help; minerals still remain in the water to clog the pads. Except in the case of sealed units, the blower motor and the blower bearings should be oiled regularly according to the manufacturer's instructions and the water reservoir flushed out as often as necessary to remove mineral deposits, sludge and scum.

Installing an evaporative cooler. Attach the mounting brackets to the sill of a double-hung window. Attach the leveling screw brackets to the bottom of the unit with their screws fully extended. With a helper, insert the outlet of the unit into the window. Hook the underside of the outlet to the window-sill brackets, and rest the plates at the heads of the leveling screws against the wall. Screw the plates to the wall. Close the window onto the unit, then back off the leveling screws until the back of the outlet grille touches the window and the cabinet is level. Close the gaps on the sides of the grille with the adjustable panels supplied with the unit.

Thread a faucet adapter (below) onto an outdoor faucet. With compression fittings (page 40, Step 2), attach ¼-inch tubing to the faucet adapter and to the threaded inlet pipe at the back of the unit. If no faucet is nearby, drill a hole near the unit through the side of the house into the basement. Run tubing from the unit to a cold-water pipe; connect the tubing to the pipe with a saddle valve and compression fittings.

BLOWER

FAUCET ADAPTER

MOUNTING BRACKETS

FLOAT VALVE

PUMP

PAD

Making the Most of Room Air Conditioners

The typical room air conditioner is a compact unit designed for quick window mounting with only a screwdriver. The only problem in installing this type is weight, which may exceed 80 pounds.

Weight becomes a more serious obstacle if you plan to use one large unit in a window *(below)* or in a through-the-wall *(pages 104-105)* opening to cool several rooms *(pages 106-107)*. A machine of sufficient capacity may weigh nearly 300 pounds and may require the use of a portable hoist, available at tool-rental establishments. Unlike smaller units, which are usually mounted with chassis and cabinet together, large air conditioners are designed so that the chassis—the heaviest part—slides out of the cabinet.

Most room air conditioners are provided with mounting angles on the top and the sides of their cabinets for use in double-hung windows; the bottom sash is then lowered onto the top of the unit to help support it. When the installation has to be made in a sliding or in a casement window, a different arrangement of mounting hardware is called for: in a sliding window or crank-operated casement window, a retractable frame rises from the cabinet to provide top support and to hold a filler board; in a fixed casement window, mounting angles on all four sides hold the unit in the window.

Mounting a Large Unit in a Double-hung Window

Preparing the cabinet. Remove the front grille and slide out the chassis. You may also have to remove a bolt that secures the compressor to the cabinet. Apply a strip of ⅜-inch adhesive-backed foam seal, available at hardware stores, under the front of the cabinet to prevent air leakage. Mounting angles are sometimes factory installed. If not, press another strip of adhesive seal over the predrilled holes and attach the top angle. Then fit the side angles into the filler-board gaskets, which seal the angles to the cabinet, and screw on the angles.

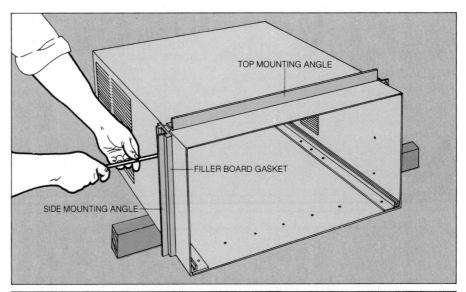

TOP MOUNTING ANGLE

FILLER BOARD GASKET

SIDE MOUNTING ANGLE

Mounting the support brackets. Units that extend more than a foot beyond the sill require exterior support brackets. A number of predrilled holes are usually provided on the brace to allow for various widths of window stools. Screw the bracket's vertical leg into the first hole beyond the stool, then fasten the braces to the stool with wood screws, ¾ inch long.

Slide the empty cabinet out until it rests on the stool and brackets, and fasten it to the braces if holes are provided. Lay a level on the cabinet, directly above each bracket, and adjust the leveling screws until the bubble in the level is half off its center line, indicating a slight downward tilt to the outside. This tilt ensures that excess condensation drips outside, not toward the house. Then fasten the cabinet to the stool with No. 10 wood screws before sliding the chassis inside and reaffixing the front grille.

Add a strip of sash gasket on the top angle before lowering the bottom sash; add filler boards at the cabinet sides and stuff sponge foam between the top and bottom sashes. You may want to affix an L-shaped window-lock bracket to keep the lower sash from being moved.

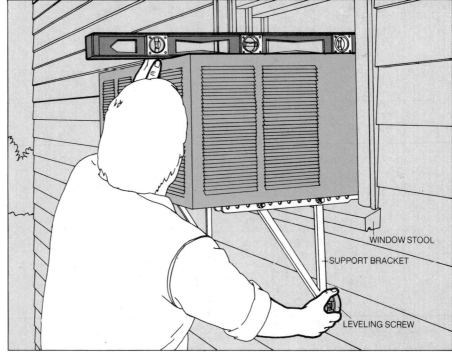

WINDOW STOOL

SUPPORT BRACKET

LEVELING SCREW

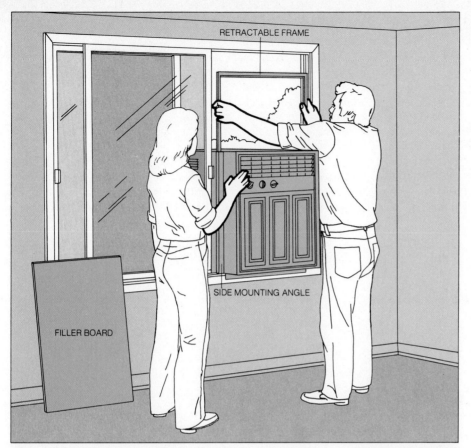

RETRACTABLE FRAME

SIDE MOUNTING ANGLE

FILLER BOARD

Special Installations for Special Windows

Sliding windows. If your windows slide open horizontally instead of vertically, or if they are the crank-open casement type, you need a specially designed air conditioner, slimmer and higher than the conventional unit. It contains a retractable frame that can be pulled upward out of the cabinet to the top of the window frame. When filled with thin filler board or transparent plastic, the frame seals off the area above the air conditioner and provides support at the top of the window frame. Where a frame is not provided, cut a piece of ⅛-inch hardboard to fit the opening.

The unit is fastened to the window frame with the side mounting angles, which may require the removal of a casement window's cranking mechanism. If this is done, be sure to wedge a piece of wood between the cabinet and the open window so that the window cannot swing against the air conditioner.

LOCKING SLIDE

SIDE CLIP

Fixed casement windows. A machine small enough to fit through a single panel is needed for a fixed casement window, preferably a unit with factory-supplied mounting hardware: retractable locking slides on both sides, mounting angles on all four sides, and side clips to hold the unit firmly without drilling holes into the casement frame. Remove glass and putty from one of the bottom panels. Slide the air conditioner through the opening, tilting the mounting clips forward if necessary to provide extra clearance. With the unit snug against the frame, adjust the slides and tighten the clip screws.

A Technique for Piercing Walls

Constructing a through-the-wall opening for a room air conditioner involves the same sort of carpentry required to build a window. For this installation you need an air conditioner long enough to extend completely through the wall so that the louvers are not obstructed at either end.

If possible, locate the opening on a shaded side of the house, close to an electrical outlet and at least 2 feet above the floor—this will increase efficiency in circulating air. Drill test holes to be sure pipes or wires will not interfere.

The opening itself is carved out in two stages. From inside, cut through the interior wallboard from floor to ceiling, exposing the studs and exterior wall. Then go outside to saw an opening the size of the air conditioner in the outside wall if it consists of wood, stucco or aluminum; chisel out the opening if the wall is of brick or other masonry construction.

1 **Opening the wall.** Mark the width of the air-conditioner cabinet on the interior wall, and extend the marks to indicate an opening just inside the studs beyond the cabinet width. Remove the baseboard and the molding from the wall. Use a circular power saw to cut the wallboard from floor to ceiling, then pry off the section. Remove any insulation.

From inside, mark an opening at the desired spot on the exterior wall, making it ½ inch higher and wider than the cabinet dimensions. Drill four small holes at the corners, then use the holes as guides to saw through, from outside, the exterior wall covering and its interior sheathing. Then saw through the studs at the opening (*right*) and pry the pieces out.

2 **Framing the opening.** From inside, nail 2-by-4s as outer and jack studs on either side of the opening. The outer studs should extend from top plate to sole plate, while the jack studs should be even with the top of the opening and far enough apart so that the empty cabinet can be slipped between them with about ¼ inch to spare.

Place a header—two 2-by-6-inch boards separated by a sheet of plywood—on top of the jack studs and nail it in place. Measure a pair of cripple studs to fit between the header and top plate, and toenail them. Toenail a double sill—two 2-by-4s laid flat—to the jack studs along the bottom of the opening, and toenail another pair of cripple studs from sill to sole plate *(right)*. Slide the cabinet into the framed opening to make sure that it slides through easily.

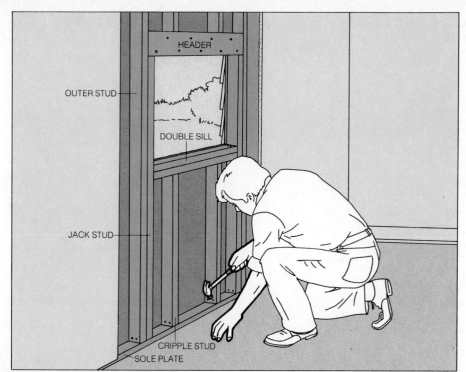

3 **Installing the cabinet.** Mount the exterior support brackets *(page 102)* and slide the empty air-conditioner cabinet through the wall and onto the brackets. Drill holes in the cabinet—two on the top and four on each side—to fasten the cabinet to the header and jack studs with No. 10 wood screws 1 inch long. Shim the gap between the bottom of the cabinet and the sill with a thin strip of wood, as shown above.

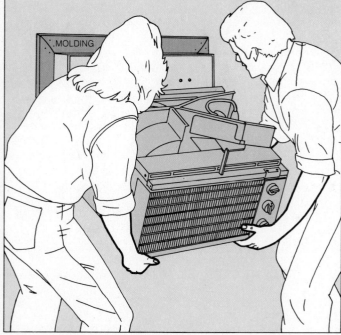

4 **Completing the installation.** Caulk the cabinet edges, both inside and outside, before replacing the wallboard above and below the cabinet. Frame the interior side of the opening with decorative molding and replace the baseboard and molding previously removed. You may also wish to frame the exterior side with molding to match that of nearby windows. Get someone to help you lift the air-conditioner chassis and slide it into the cabinet *(above)*. Then attach the front grille and plug in the unit.

Double Duty from Room Units

Many people use one large window-mounted air conditioner to cool several rooms or even a whole house. It works, provided you get a unit of sufficient capacity and can arrange for cold air to circulate freely. This approach is almost always less expensive than central air conditioning, especially in a house without heating ducts.

If you plan to use a window unit to cool more than one room, its placement is crucial. To serve two rooms separated by conventional walls and doorways, the best position is on a wall directly facing the doorway of the second room. The technique is more effective—especially where more than two rooms are to be cooled—when auxiliary fans are used to force the cool air through doorways or around sharp turns (below).

If you plan to cool an entire floor with a window unit, there are two good locations for it. One is at the end of a central hallway (opposite, top). The second is in a room located near the center of the floor, which reduces the distance the cool air must travel.

In a house with a forced-air furnace, there is a third choice if you can place the air conditioner so that it blows cold air toward a return register of the heating system. Turn on the furnace blower (page 36) to circulate cool air through the heating ducts to every room in the house. This measure is even more effective if you increase the blower speed during the cooling season (pages 37 and 38).

If cool air must be forced past more than two doorways or around more than two sharp corners, as in the two-story house opposite, bottom, consider a second air conditioner. Even if your house requires three medium-sized units to cool it effectively, they will generally cost less to install and operate than a central air-conditioning system.

Once you have determined appropriate locations for the air conditioners, use the rules on page 125 to calculate the size units you require. But add an extra 10 per cent to the floor area that each unit must cool to compensate for the loss in cooling efficiency caused by the restrictions in air circulation.

As versatile and economical as window air conditioners can be, they are less efficient in most ways than properly installed central units. A large window air conditioner running at top speed is necessary to cool an entire floor. The unit is likely to be noisy, and the space nearby may be too cold for comfort. You can reduce such side effects by installing the unit away from the room to be cooled, and using fans—which are fairly quiet—to bring in the cool air.

Directing cold air. A portable fan set in a doorway (left) pushes air from an air-conditioned room into an adjacent one. Use a 16-inch floor model; it can circulate a large volume of air without unpleasant drafts or excessive noise. Alternatively, install a smaller fan—a kitchen exhaust fan, for example—through the wall above the door. A third solution is to place a small exhaust fan in a window to pull the cool air (right)—but it exhausts some cool air from the house, forcing the air conditioner to work harder.

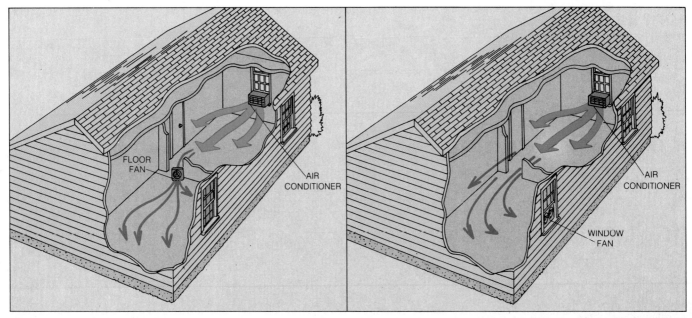

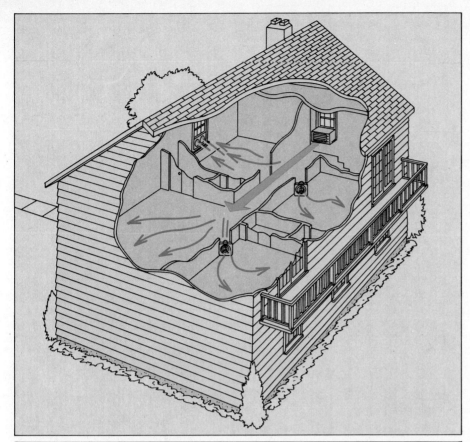

Air-conditioning several rooms. A large air conditioner at one end of a central hallway can cool an entire floor. The unit blows some cool air into the room at the opposite end of the hall, but fans are necessary to divert the air stream into the other three rooms. One room has a conventional window with an exhaust fan to pull cool air; the other rooms, which have French windows, are cooled by air pushed in by floor fans in the doorways to the hall. By closing doors of some rooms and turning off fans, cool air can be channeled to where it is needed most.

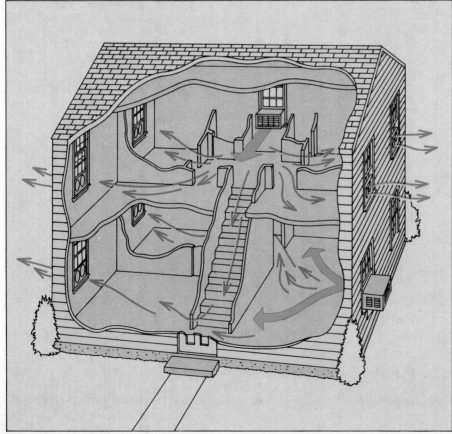

Cooling two floors. Two air conditioners are usually necessary to cool a two-story house—one upstairs and one down. In the example at left, air conditioners cool the upstairs hallway and a large room on the first floor. Some of the cool air is distributed among the rooms upstairs by fans; the rest of it flows by gravity down the stairway to the first floor. The air that goes downstairs, which can be regulated by closing off rooms and changing the air-conditioner setting, is mixed by a window fan with air from the first-floor air conditioner to cool the downstairs room in the foreground. The downstairs room in the background, also fitted with a window fan, gets most of its cool air from the downstairs unit.

Adding Air Conditioning to a Forced-Air System

If you plan to add central air conditioning to a house with steam or hot-water heat, you will have to install a network of supply and return ducts to circulate cool air throughout the house *(pages 118-119)*. But if you have a forced warm-air furnace, the existing blower and ductwork can usually be adapted for air conditioning with minimum modification.

The key to this transformation is "split system" central air conditioning *(opposite)*, in which the cooling coil is mounted in the furnace plenum while the other components are placed outdoors. In older installations, both indoor and outdoor components and their connecting tubing had to be filled, or charged, with refrigerant gas after they were in place—a job that required professional equipment and expertise. The more recent development of factory-charged components and quick-connect couplings for the tubing *(page 113)* has greatly simplified the job, and several manufacturers and large retail stores now market the precharged units in kits *(right)* for homeowner installation.

The installation involves only about two days of routine work with sheet metal, tubing and wiring. However, before beginning, check your building department about local regulations governing the placement of the condensing unit and the installation of new circuits—some wiring for central air conditioning may pose problems for which you may want professional help *(pages 115-117)*.

Before ordering the installation kit, determine how large a system is needed to cool your home and make sure that the ductwork, blower and electrical service are adequate for air conditioning. You can make this estimate by using the data on pages 124-125, but you may want to ask your dealer to confirm your figures. Correct sizing of the system is important: too small a unit will not cool adequately on very hot days, too large a unit will not run long enough to lower humidity.

Air conditioning puts much more severe demands on the air-circulating system than heating for two reasons. First, while furnaces can readily raise air temperatures by as much as 100°, cooling coils can efficiently lower them by only about 10° to 20°; to offset the smaller temperature difference, a larger volume of cold air must be moved to cool the same area. Second, since cold air is heavier than warm air, more power is needed to force it through the duct system.

To meet these requirements, a higher blower speed—and, in some cases, a more powerful blower motor—is necessary to circulate the cooled air. Blower speeds can often be increased by a simple wiring change *(page 38)* or by a pulley adjustment *(page 37)*. But if the blower motor becomes overloaded and shuts off at the higher speed, you will need a more powerful one.

To provide better air flow, you also can modify or expand the existing ductwork. Some alterations, which apply to both single- and two-story homes, are very simple. Adjust the dampers *(pages 10-11)* to send most of the cold air upstairs or to the areas most exposed to the sun's heat. To prevent cool air from pooling near the floor, adjust the supply registers or attach plastic air deflectors to direct incoming air upward. It may also help to replace registers with a type having directional grilles designed for cold air.

Additional measures may be necessary to keep a second floor comfortable, particularly if the attic is not heavily insulated or well ventilated. You may have to add ducts to supply more cold air, and returns to bring down hot air from near the ceilings. Ducts can often be installed fairly simply as shown on pages 54-59.

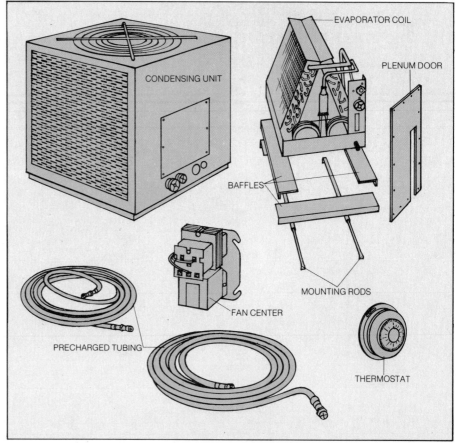

A kit for central air conditioning. In addition to an outdoor condensing unit, an indoor evaporator coil and precharged tubing to connect them, a typical kit includes mounting rods for the evaporator coil, precut baffles to seal the space between the coil and the plenum, a precut metal door with cutouts for the refrigerant and drain lines, a heating-cooling thermostat, and a transformer-fan relay unit called a fan center for the low-voltage wiring. You will also need caulking compound, pipe hangers or clamps for the tubing, piping for the evaporator drain, 24-volt cable, and a cutoff switch and cable for the 240-volt condensing unit circuit.

Where the pieces fit. When installed in a typical warm-air heating system the components on the opposite page transform the ductwork, filter and blower into a central cooling system. The condensing unit sits outside on a concrete base and is connected to the evaporator coil, mounted on the furnace, by a small liquid refrigerant line and a larger gas, or suction, line. In operation, the furnace blower pulls warm air through the return ducts, filters it and circulates it through the cooling coil in the plenum to the supply ducts. Humidity removed in the cooling process runs down the coil to a condensate tray and then through a drain line to a floor drain or laundry tub. A separate 240-volt circuit, with an outdoor safety switch, supplies the power needs of the compressor motor and fan in the condensing unit; indoors, low-voltage wires relay signals from the thermostat through the fan center mounted at the furnace, turning on the blower motor and the compressor when cooling is called for.

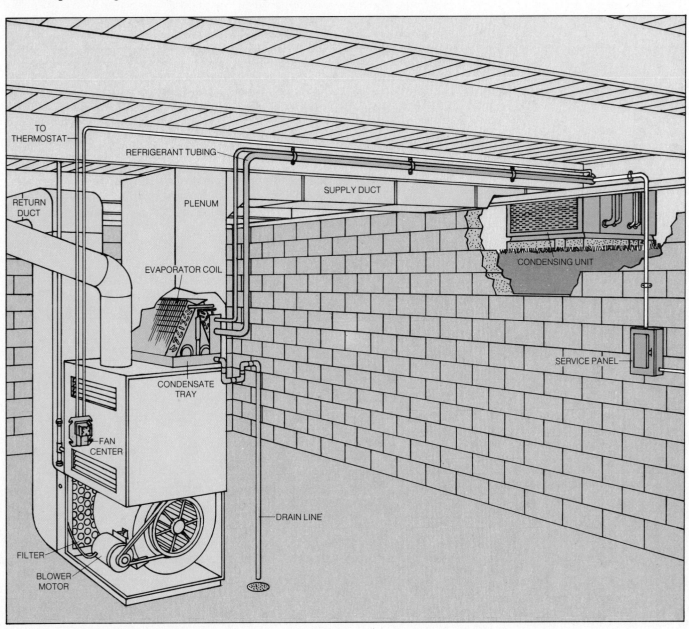

TO THERMOSTAT

REFRIGERANT TUBING

SUPPLY DUCT

RETURN DUCT

PLENUM

CONDENSING UNIT

EVAPORATOR COIL

SERVICE PANEL

CONDENSATE TRAY

FAN CENTER

DRAIN LINE

FILTER

BLOWER MOTOR

Hooking Up the Cooling Components

Once you have determined the capacity of the units required to cool your home, obtain the dimensions of the indoor and outdoor sections from the manufacturer or dealer. You can then prepare a base for the heavy condensing unit before it arrives, allowing you to unload it directly into its permanent position.

To locate it properly, determine the maximum length of precharged tubing available for the recommended units. Within that limit, place the outdoor section as close as possible to the furnace but where its noise will be least annoying to your neighbors as well as yourself.

(Some communities have codes specifying a minimum distance between the noisy condensing units and the property line.) Leave at least 18 inches of clearance between the unit and the foundation wall. Make sure there is unrestricted air flow to the intake grilles, and that hot air does not discharge directly onto the house or onto flowers and shrubbery.

Indoors, measure width, depth and height of the furnace plenum to make sure the evaporator coil will fit inside. The bottom of the coil must be high enough so that air can flow freely past the furnace heat exchangers underneath—a 3- to 4-inch clearance is usually adequate—while the top of the coil must not project more than 2 inches above the bottom of the lowest duct; above that, the coil would block air flow to that duct.

If your plenum is too small, you will have to replace it with a larger unit matched to your furnace and duct sizes.

In the installation shown opposite, bottom right, the coil is supported on adjustable steel rods that rest on top of the furnace flange. In some systems *(inset)* metal angles screwed to the plenum sides support the coil. Moisture condensing on the coil must be removed by a drain line *(page 112);* if there is no drain near the furnace, you may have to install a condensate pump *(page 120)* to dispose of the water.

Use wire or string to estimate the length of refrigerant tubing needed; precharged tubing is available in lengths from 10 to 45 feet, increasing in 5-foot increments. Be careful not to kink or crimp the tubing and, before tightening the couplings *(page 114)*, make sure that they are free of dust and dirt, properly lubricated and correctly threaded. If you make a mistake in screwing connectors together you cannot correct it; once joined, the connectors cannot be opened without releasing the refrigerant gas.

Locating the Condenser

1 Preparing the base. At the location chosen for the outdoor unit, level about 9 square feet of ground and spread a thin layer of sand over it. Set four solid 4-by-8-by-16-inch concrete blocks in the sand to form a base approximately 30 inches on each side. (A 4-inch-thick precast concrete slab of the same dimensions also makes an excellent base.) Make sure the base is level.

2 Positioning the unit. If the kit includes rubber washers to dampen vibration between the unit and the base, attach them to the bottom corners of the condensing unit. Use a hand truck to wheel the unit into position over the blocks. Then, with a helper, lower it onto the base *(right)*, making sure that the refrigerant couplings are adjacent to the foundation wall. Directly opposite the couplings, drill a ¼-inch pilot hole just above the sill plate; then enlarge the opening to a 3-inch diameter with a hole-saw attachment.

Inserting the Evaporator Coil

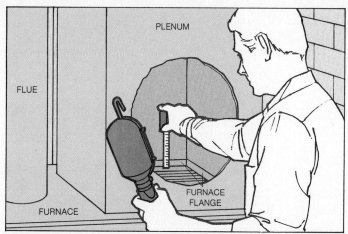

1 Measuring the furnace flange. After shutting off the electrical supply to the furnace, remove the flue if it will obstruct the coil installation. Wearing gloves and long sleeves to protect against sharp metal edges, open a 10-inch diameter hole in the plenum (*page 51*) approximately 12 inches above the furnace top. Through the hole, measure the height of the furnace flange to which the plenum is attached. Add ¼ inch to that measurement, and, on the front of the plenum, scratch a horizontal mark that distance from the furnace top (use a level as a guide for an awl). Make a second horizontal mark on the plenum ½ inch below the first mark.

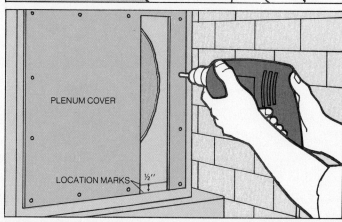

2 Making the cover holes. Align the bottom of the precut cover with the second location mark (*Step 1, above*), making sure that the slot cutout corresponds to the location of the refrigerant-tube couplings on the evaporator coil. Mark the location of the cover holes at top right and bottom left, drill pilot holes and attach the cover with sheet-metal screws. Using the cover as a template, drill the remaining cover holes and complete marking the cover outline on the plenum.

3 Completing the plenum opening. Remove the cover and mark a line ½ inch inside the cover outline (*Step 2, above*). Use tin snips to cut away the plenum along the inner line. Double-check the dimensions of the opening; it should be about 1 inch wider and higher than the width and height of the evaporator coil.

4 Installing the coil supports. To assemble rod coil supports, slide the smaller diameter rod inside the larger. Inside the plenum, hook the flat end of the support between the rear furnace flange and the plenum; then adjust the support to the plenum depth and secure the other end between the front flange and plenum. Install the second support parallel to the first, making sure that the distance between supports is less than the width of the evaporator coil they will hold. If you use two metal angles as coil support

(*inset*), tape a torpedo level to an angle top. Hold the angle level, 4 inches from the furnace top, and use it as a template to mark the inside of the plenum for screw holes. Drill the holes and secure the angle with sheet-metal screws. Install the other angle opposite the first, at the same distance from the furnace top.

5 **Cutting the baffles to size.** Cut the four sheet-metal baffles from the kit *(page 108)* so that, when installed around the plenum edges, they will form an opening the same size as the coil air intake. (If you are not using a kit, cut baffles from 24- or 26-gauge sheet metal and form a flange ⅝ inch wide along one edge, following the directions for working with sheet metal on page 51.) The side baffles are of identical size, so that the opening is centered across the plenum width; the rear baffle must be bigger than the front one to position fittings on the coil front to protrude through the plenum door.

6 **Installing the evaporator coil.** Install the front and rear baffles first, pushing their flanges down between the plenum and furnace flanges until the baffle rests evenly on the coil supports. Then install the side baffles. Press caulking compound into the joint between baffle edges and plenum sides. Lay a second bead of caulk ½ inch from the inner baffle edges. Carefully lower the coil into position over the baffle opening *(below)* with its front edge flush with the plenum front. Press down on the coil to seal the inner bead of caulk to the underside of the coil. Secure the plenum door with sheet-metal screws.

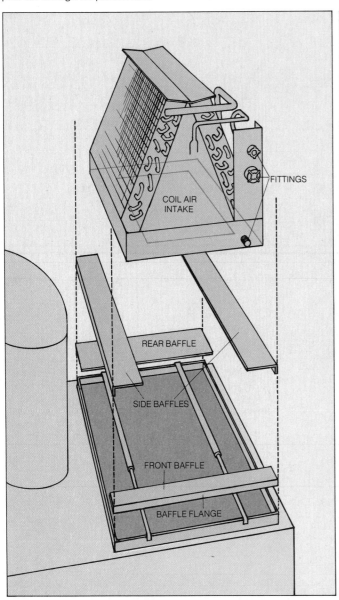

7 **Attaching a drain line.** Make a trap from elbow fittings and pipe ¾ inch in diameter (or use a ready-made trap if one is available), and attach it to the threaded fitting on the condensate tray. Connect pipe to the trap to make a drain emptying into a laundry sink or a floor drain.

Attaching Precharged Tubing

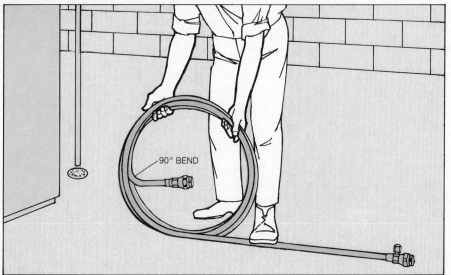

1 **Uncoiling the tubing.** Hold the coil of suction tubing in a vertical position and place your foot lightly over the straight end. Moving your foot as necessary to hold the straight end level with the basement floor, unroll the tubing so that the other end, which has a 90° bend, uncoils near the evaporator coil. Unroll the liquid line in the same way. Feed the straight end of each tube through the hole in the outside wall (*page 110, Step 2*) to the condensing unit outdoors.

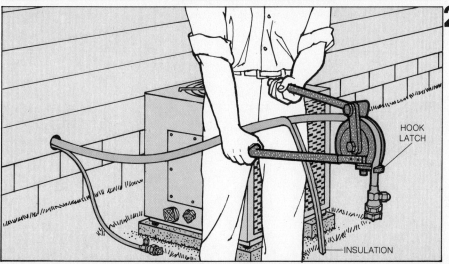

2 **Forming bends.** Gentle bends in the refrigerant lines can be formed by hand, but for sharp bends, try to borrow a tube bender that fits the tubing diameter; many equipment dealers lend them. To use a bender, peel insulation from the tubing section. Insert the tubing into the bender, close the hook latch and press the handles until the required shape is achieved. Replace the insulation and patch the slit with duct tape.

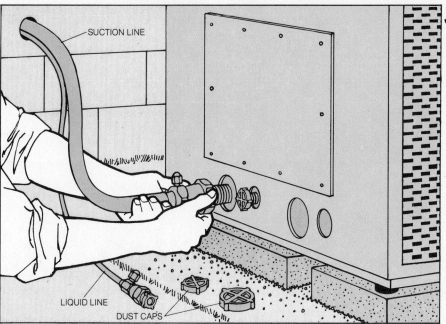

3 **Connecting the condensing unit.** Unscrew the dust caps from the suction line and the matching fitting on the outdoor unit. Make sure the threads are clean and that a rubber O-ring is in place inside the condenser fitting. Oil threads of both fittings with refrigerant oil, available at refrigeration-equipment dealers. Holding the outer hex nut steady, carefully align the threads and turn the inner hex nut approximately two full turns onto the threaded fitting. Repeat this procedure for the liquid line, hand-tightening its coupling approximately three turns. Do not tighten the couplings any further until the rest of the tubing has been formed and routed.

4 **Routing the tubing.** Indoors, use duct tape to tie liquid and suction lines together about every 4 feet. Then route the tubing along a joist from the outside wall to the furnace, loosely fastening it with large cable clamps. If you have several feet of excess tubing at the furnace, loosen the clamps and hand-form a large horizontal loop. (A vertical loop could trap oil or liquid refrigerant, which might damage the compressor.) Clamp the loop to joists and clamp the remainder of the tubing every 4 feet. Over the furnace plenum, make a wide bend to align the ends of the tubing with the liquid and suction fittings on the front of the evaporator coil.

FURNACE PLENUM

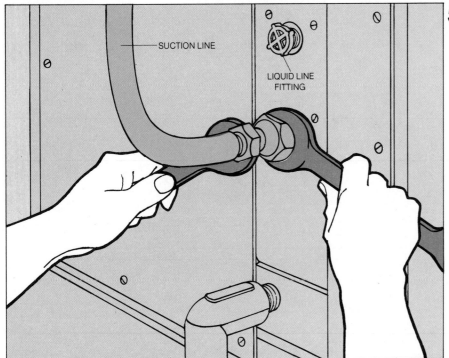

SUCTION LINE

LIQUID LINE FITTING

5 **Connecting the coil.** At the evaporator coil, lubricate the threads and hand-tighten the suction line, following the procedure in Step 3, page 113. Continue tightening the coupling with wrenches, holding the outer hex nut steady while turning the inner nut clockwise. You will hear a hissing sound as the diaphragms that seal in refrigerant gas are pierced; keep tightening until you feel a marked increase in resistance as the internal brass seals make contact. Then tighten an additional quarter turn and stop. Attach the liquid line in the same way and coat the connections with liquid soap to check for leaks. (Should you find one, tighten the nut another eighth of a turn and, if the leak persists, notify the dealer.)

Once the evaporator connections are satisfactory, go outside and use wrenches, as described above, to complete tightening those ends of the tubing—suction line first. Check for leaks.

Making Electrical Connections

Wiring a central air conditioner involves working with high-voltage lines that supply power to the compressor motor and low-voltage wires that relay signals from the thermostat to the other components. Unless you have had considerable experience in reading low-voltage wiring diagrams and connecting 240-volt lines to the service panel, you may want to leave these parts of the job to a professional familiar with heating-cooling installations. However, running 240-volt cables and connecting them to the safety switch and air-conditioning compressor involves only basic skills, and you can save considerable expense by doing this part of the work yourself.

To install a 240-volt circuit to the compressor, you will need two-wire weatherproof cable with a ground wire—No. 10 unless the compressor name plate indicates an amperage exceeding 30 amperes, a load that requires No. 8. In addition, a safety switch must be installed near the outdoor unit so that power can be cut off in an emergency or for servicing. Check whether your unit has one built in near the access panel; if not, you will need to mount a 240-volt outdoor safety switch on the wall near the condensing unit.

To control the new cooling equipment,

the existing heating thermostat must be changed to a heating-cooling unit. On some thermostats, the change simply involves adding a special base to the existing model. The low-voltage power, on which most thermostats operate, is provided by a 24-volt transformer, usually mounted on a junction box in the furnace compartment.

Many newer furnaces have the transformer and all the terminals necessary for the later addition of cooling equipment incorporated into a unit called a fan center; if your furnace has a fan center, the low-voltage connections are fairly simple. However, on many older furnaces you may find only heating transformers, which cannot accommodate the additional load of air conditioning.

If your furnace has only a heating transformer, replacing it with a fan center and making other connections is complicated—the existing wires must be traced on the wiring diagram for the furnace, then modified to accept the air conditioner. The wiring changes required to add a fan center to a typical gas furnace are shown on pages 116-117. However, connections vary considerably on different furnace types and models, and you may need professional assistance.

Connecting the condensing unit. Outdoors, mount a weatherproof 240-volt safety switch that matches to the amperage specified on the unit name plate. Using the hole for the refrigerant tubing, run a cable to the switch from the service panel. Connect the two insulated wires to the terminals marked LINE (*inset, left*), wrapping black tape around the white-insulated wire to recode it as voltage-carrying. Fasten the bare wire to a green grounding screw.

Run cable from the switch to the condensing unit, fastening it to the refrigerant tubing every 2 feet with cable straps. Connect the insulated wires to the LOAD terminals on the switch, recoding the white wire. At the condenser, remove the access panel and connect the insulated wires to the two line terminals, marked L1 and L2 (*inset, right*), recoding the white wire. Ground the bare wire at switch and condenser. The low-voltage wires are attached to the terminal block at the factory, but must be connected to the fan-center controls during installation (*page 117*).

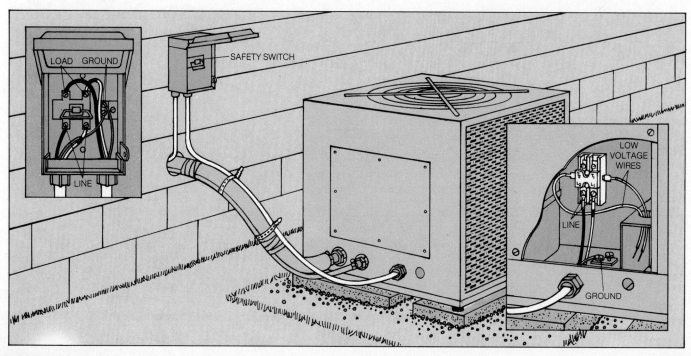

Removing the heating transformer. On a typical, fairly new gas furnace, shut off power at the furnace switch, usually mounted near the point where the power cable enters the furnace. Remove the heating thermostat (*page 16*) and check the color coding of the wires attached to the terminals marked R and W. At the furnace, remove the cover from the furnace compartment and trace the cable containing the two thermostat wires. Disconnect one wire from the gas valve terminal marked TH and the other from the transformer terminal marked R. Loosen the wire from the other gas valve terminal marked TR at the transformer's C terminal.

With all the low-voltage wires disconnected, loosen the screws holding the transformer mounting plate to the furnace junction box. Inside the box, disconnect the transformer's two 120-volt leads—the white lead from the other neutral wires and the black, or hot, lead from the wire to the limit switch. Discard the transformer.

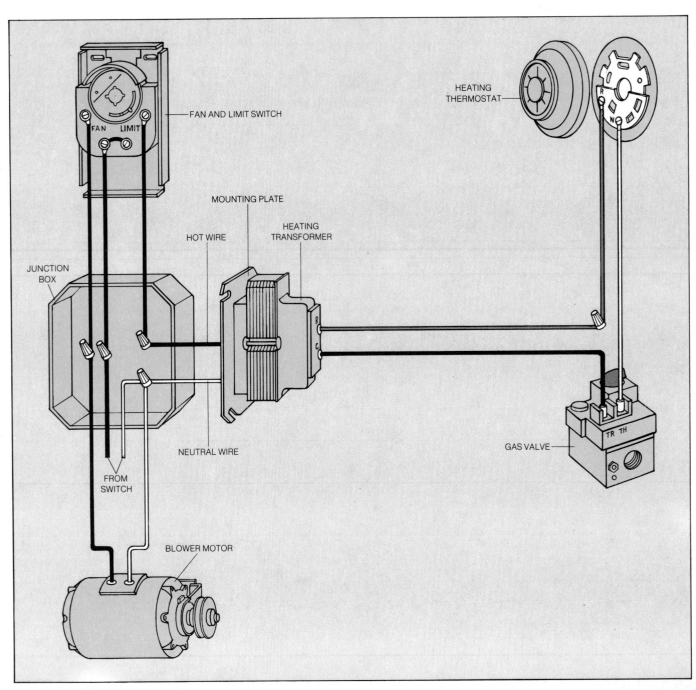

FAN AND LIMIT SWITCH

HEATING THERMOSTAT

MOUNTING PLATE

HOT WIRE

HEATING TRANSFORMER

JUNCTION BOX

NEUTRAL WIRE

FROM SWITCH

GAS VALVE

BLOWER MOTOR

Installing the fan center. Wiring varies widely, but the following instructions apply to many furnaces now in use. At the fan center mounting plate, separate the black and white leads marked PRIMARY from the two black fan relay leads. Use wire caps to connect the black primary lead to the limit-switch wire disconnected earlier *(left)* and the white primary to the two neutral wires in the junction box. Join either of the two fan relay leads to the hot wire running to the fan and limit switch.

Connect the other fan relay lead to the wire from the fan switch to the blower motor. Secure the fan center's mounting plate to the junction box.

At the thermostat, install a new base or, if necessary, a heating-cooling thermostat. Pull out the old two-wire thermostat cable, using it to pull new four-wire cable—color-coded red, green, white and yellow—into position at the same time. Connect the color-coded wires to the corresponding terminals on the thermostat base: red to R, green to G, white to W and yellow to Y.

At the furnace, connect the other ends of the cable wires to the screw terminals marked R, G, W and Y on the front of the fan center. Attach the wire from the gas valve's TR terminal to the C terminal on the fan center and run another length of wire from the gas valve's TH terminal to the W terminal on the fan center.

Run a two-wire 24-volt cable from the furnace to the outdoor unit, using the hole for the refrigerant tubing and high-voltage cable. In the furnace, connect one wire to the fan center's C terminal and the other to the Y terminal. Outdoors, remove the access panel from the condensing unit *(page 115)* and locate the two low-voltage leads. Using wire caps, join the fan center's C terminal to the black lead and the Y terminal to the yellow lead. Replace the access panel and fill the wall hole with butyl rubber caulking.

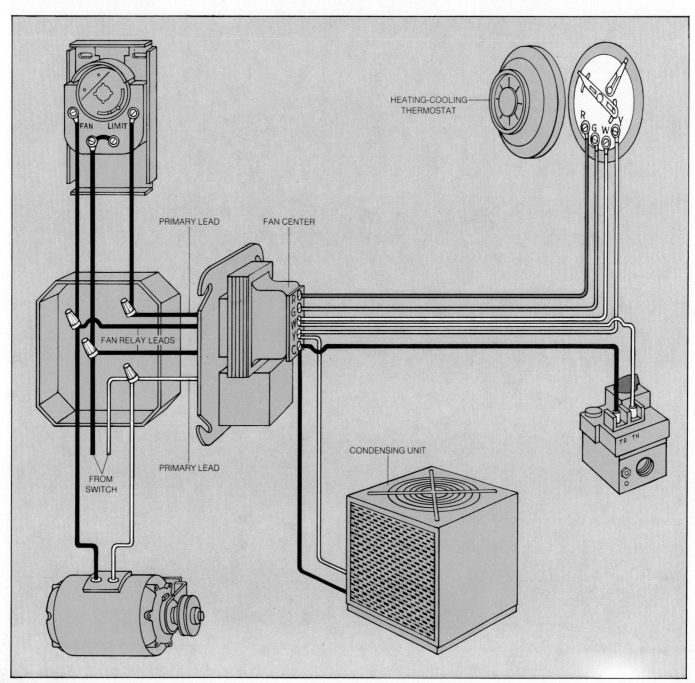

FAN LIMIT

HEATING-COOLING THERMOSTAT

R G W Y

PRIMARY LEAD

FAN CENTER

R
G
W
Y
C

FAN RELAY LEADS

PRIMARY LEAD

FROM SWITCH

CONDENSING UNIT

TR TH

Air Conditioning from the Attic

You may not be able to install central air conditioning through your heating system *(pages 108-117)*—either because the ducts of your forced-air system are too small or because you have water or steam heat or no heat at all. In that case, you can cool your house just as effectively from the attic, with ducts and a blower to distribute the cold air. The noisy condensing unit goes outdoors as usual *(page 110)*, and the blower unit, which comes with an evaporator coil already installed, usually goes into the attic. From there, the ceiling of each top-floor room—the ideal location for a cold-air outlet—is easily accessible.

If you are air-conditioning a two-story house, run ducts to the first floor through second-floor closets or box them into corners of rooms, as shown on pages 58-59. If the house is very large, consider a dual system with two condensing units outdoors, and one blower unit in the attic to cool the upstairs and another in the basement to cool the ground floor.

You can determine yourself what size air conditioner you need to cool your house *(pages 124-125)*, but have the contractor or dealer who sells you the unit tell you the size, number and placement of ducts for the best cooling effect. Three of the sheet-metal components—a supply plenum, a return plenum and an emergency drain pan to catch any overflow of condensate from the cooling coil—must be custom-made to the contractor's specifications. The ducts, grilles and registers can be obtained at heating-and-cooling-supply stores. Use an adjustable register or a damper in each supply duct. Cover the return plenum with a grille that holds a filter. This filter takes the place of one in the blower so that you can change filters without climbing up to the attic or down to the basement.

Finally, blowers and ducts installed in attics must be insulated from summer heat and humidity. Ask for a blower unit that is factory insulated. The sheet-metal shop that makes your plenums will insulate them for you. You can use flexible, insulated ducts *(page 54)* to carry the cold air, but if runs are long it may be more economical to install rigid duct and wrap it with special duct insulation, vapor barrier facing out.

An attic installation. In a typical single-story house *(below)*, a blower unit containing an evaporator coil sits in an emergency drain pan laid across joists. Round supply ducts lead from a supply plenum, screwed to one end of the blower, to ceiling registers in each room. A return plenum extends from the blower intake into a central hallway below. Drain pipes from the evaporator coil and emergency drain pan, as well as refrigerant lines to the condensing unit, run outdoors at the eaves. Such a simple setup may even cool some two-story houses, since chilled air will flow down the stairwell, but for more efficient operation additional ducts like those on pages 54-59 may have to be installed.

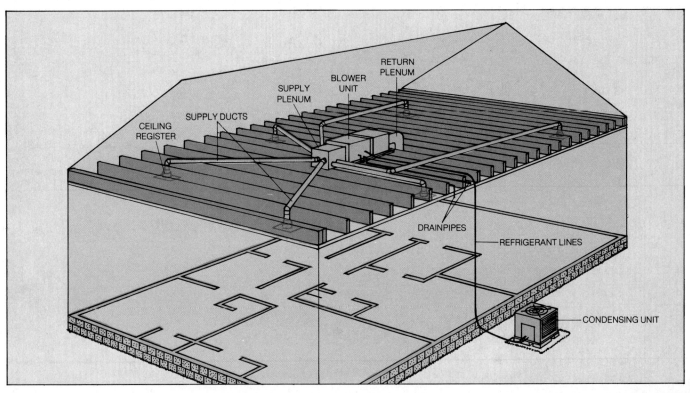

Installing the blower and ducts. Set the drain pan on the joists. To raise the blower above any overflow water, place 2-by-4s treated with a wood preservative in the pan. Set rubber cushions on the 2-by-4s at each corner to absorb vibration from the blower unit. Hoist the unit into the attic with a block and tackle suspended from the rafters. With a helper, lower the unit onto the rubber cushions in the pan (*below, left*).

Screw one end of the return plenum to flanges at the blower intake and rest the other end on the ceiling of the floor below, against the joist. You will have to cut a joist, using the method shown on pages 98-99, if the plenum is wider than the joist spacing. Next, cut a hole in the ceiling for the plenum, lower the end of the plenum through the hole and bend the flanges to lie flush against the ceiling (*below, right*). Drill

holes through the flanges, and fasten them to the joist and ceiling with screws and toggle bolts. Then push the filter grille into the plenum and screw it to the plenum flanges. Install a filter and screw the cover shut.

In the attic, screw the supply plenum to the other end of the blower, then install ducts to each room as shown on pages 54-59.

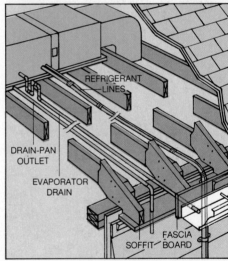

Drains and refrigerant plumbing. In the attic, measure the distance from the outlet on the emergency drain pan and from the drain and refrigerant line connections on the blower unit to an adjacent joist. Outside, use the nails holding the soffit to the joists or rafters to help locate the joist. Using the measurements taken inside, drill a 1-inch hole in the soffit for the emergency drain and a 3-inch hole for the refrigerant lines and electrical wiring. Drill a 1-inch hole through the fascia board above the gutter—or through the side if necessary (*above*)—for the evaporator drainpipe. Assemble drain lines of ¾-inch pipe. Include a trap in the evaporator drain, and pitch the pipe away from the blower unit so it will drip into the gutter. Run the emergency drain through the soffit so that any overflow will be visible. Install the refrigerant lines, using the techniques on pages 113-114.

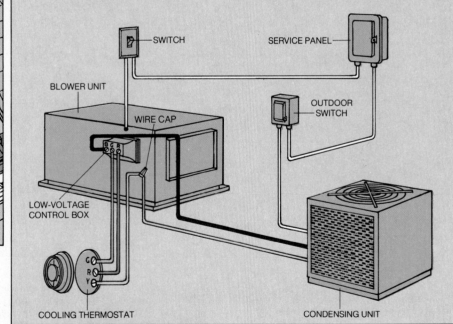

Wiring. Choose a location for the thermostat (*page 16*) and drill a ½-inch hole. Fish a three-wire, No. 18 cable through the hole to the blower unit and a two-wire cable along the route of the refrigerant lines from the blower to the condensing unit. Connect the three-wire cable to the terminals marked Y, R and G on the thermostat wall plate, and write down what color wire you attached to each terminal.

At the other end of the cable, connect the wires from the R and G thermostat terminals to the R and G terminals in the low-voltage control box

on the blower unit. With a wire cap, connect the third wire to one of the wires in the two-wire cable. The second wire goes to the B terminal in the control box. At the condensing unit, connect the first wire of the two-wire cable to the Y terminal or yellow wire of the condensing unit and the second to the B terminal or black wire.

High-voltage wiring consists of a 240-volt line from the service panel through an outdoor cutoff switch to the condensing unit (*page 115*) and, depending on power requirements, a 120- or a 240-volt line through a switch to the blower.

Repairs for Air Conditioners

Room air conditioners, central air conditioners and heat pumps—simply reversible air conditioners *(page 122)*—are remarkably trouble-free. The principal maintenance needed is cleaning. Replace disposable filters and wash reusable ones about once a month. Owners of older units that are not permanently lubricated should oil fan and blower motors and bearings once a year.

Meticulous homeowners, aware that dirt on refrigerant coils and fans reduces efficiency, start each season by cleaning the air conditioner. More often, however, no attention is paid an air conditioner or a heat pump until something goes wrong.

As the troubleshooting chart overleaf shows, there are minor ailments you can cure yourself. Dirty components are often the problem, and cleaning restores efficient operation. The more serious difficulties in a central unit most often occur in the thermostat and blower—the parts it shares with the heating system.

Before repairing an air conditioner or a heat pump, unplug it or turn off its electrical power at the service panel. Then remove the access panels that cover the interior components. A central unit has a cover for the evaporator, or A-coil, and another for the blower, both of which are in the furnace. Outdoors, the grilles over the condenser either unsnap or unscrew. Room air conditioners may have similar panels, but many large units can be slid partway out of their casings and into the room *(page 105)* for easy access.

Cleaning an air conditioner or a heat pump requires both a gentle touch and a firm hand. Fins on the cooling coils are delicate; for them, a soft brush and a vacuum cleaner are the best tools. Fans and blowers, however, become encrusted with hard-to-remove dirt; you may need to use a wire brush or putty knife.

Before switching the unit on after you have repaired it, be sure it has been off at least five minutes; unless you give it time to dissipate excess pressure you may overload the compressor. Conversely, leaving a central air conditioner or heat pump (not a window unit) disconnected from all power for a considerable time—three days or more—also causes a problem: refrigerant in the compressor may liquefy, and in this state it can damage a working compressor. To avoid such harm, wait 24 hours after you restore power to the unit before turning on the operating switch. During this waiting period, an automatic heater in the unit will vaporize any refrigerant that has liquefied.

Cleaning the condensate box and pump. Unplug the pump motor, and take off the box cover. Remove the screws at the base of the pump impeller shaft, and lift out the pump, complete with its motor switch and float assembly. Remove the motor housing screws and lift off the housing with the switch-float assembly attached. Clean the fan on top of the motor with a brush and a kerosene-soaked rag *(below, right)*, and oil the motor, following the directions that are on the plate attached to its cover.

Scrub the inside of the box, then scour the float rod so it can move freely. After you replace the motor housing and mount the pump on its base, check that the float ball and rod do not bind, and the power cord does not dip below the motor.

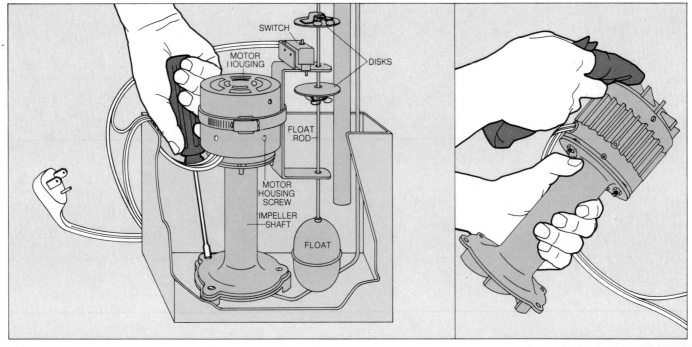

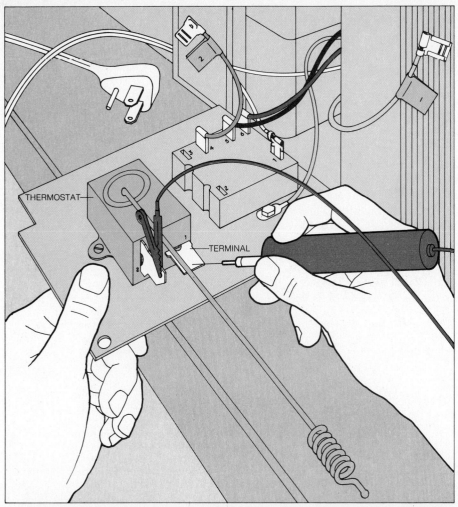

Testing a window-unit thermostat. Unplug the air conditioner and unfasten the control panel. With masking tape, label each of the two thermostat wires with the number next to its terminal *(left)*. Remove the wires and attach the alligator clip of a continuity tester to one terminal. With the air temperature near the unit between 70° and 80° and the temperature control at its warmest setting, touch the tester probe to the other terminal; the tester bulb should not light. Turn the thermostat to its coldest setting and repeat the test; the bulb should glow. If the thermostat fails either test, replace it with an exact duplicate and connect the two wires. Reattach the control panel; if the thermostat has a sensor bulb, position the new bulb in the air stream exactly where the old one was.

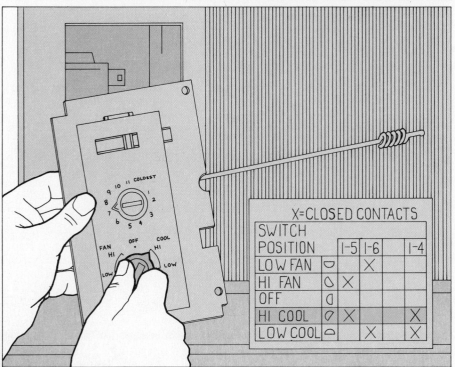

Testing a selector switch. Unplug the unit and unfasten the control panel. Pull the wires from the switch terminals, and label each one.

Find the wiring diagram—it will be somewhere inside the unit—and note the small chart *(inset)* showing terminal numbers and switch-shaft positions for each switch setting. At each setting, test all the terminal combinations shown in the chart—1-5, 1-6 and 1-4 in this example—by attaching the alligator clip of a continuity tester to terminal No. 1 and touching the tester probe to each of the other terminals. Combinations marked by Xs or a similar symbol should light the tester. With the switch shown here set at HI COOL, the tester should light for the combinations 1-5 and 1-4, but not 1-6. If the tester lights when it should not or fails to light when it should, replace the switch.

X=CLOSED CONTACTS

SWITCH POSITION		1-5	1-6	1-4
LOW FAN	▽		X	
HI FAN	◠	X		
OFF	◖			
HI COOL	◗	X		X
LOW COOL	◗		X	X

A Unit that Heats and Cools

A heat pump is essentially an air conditioner that not only draws heat from indoor air and transfers it outdoors, like other air conditioners, but can also reverse itself by means of a special valve, and draw heat from the outdoor air to the interior of a house (even cold outdoor air contains much heat energy).

Heat pumps have the same servicing needs as air conditioners and a few special needs of their own too. Almost all have mechanisms that permit them to operate under freezing conditions and that sometimes need attention.

When outside temperatures approach 32°, the outdoor coil of a heat pump frosts up like the freezer compartment of an old-fashioned refrigerator as water vapor in the atmosphere condenses on the coil. An icy sheath blocks the flow of outdoor air through the coil and prevents the coil from picking up enough warmth from the air for transmission to the house. If the heat pump is functioning properly it will defrost itself by switching automatically from the heating to the cooling mode; then heat from the house is transferred to the refrigerant, which warms and defrosts the outside coil.

At the same time, the unit automatically switches on a supplemental electric-resistance heating unit inside the house to provide warmth there until the heat pump has defrosted itself and gone back to work. Occasionally it may fail to defrost or, once started, may fail to stop defrosting. In either case, the supplemental heater keeps working, and unless you notice and correct the malfunction promptly you may incur an unexpectedly large electric bill from the continued running of that heater.

The signs of such trouble are visible on the outdoor coil. It should show frost, at least periodically, when the outdoor temperature is near or below 32°; if it stays warm and dry, the unit is defrosting continuously and you should call a serviceman. On the other hand, you may find the coil covered with a heavy coat of ice; if it shows no sign of defrosting, try setting the thermostat below room temperature to shift the unit into the cooling mode. (Some units have an additional switch you must set to select cooling.) After 10 minutes, switch the thermostat to emergency heat for 10 minutes, to keep the house warm, then back to cooling. If your problem is a sticky valve this operation may cure it.

If the unit does not defrost within an hour or so, call a serviceman. If the unit does defrost but immediately ices up again, try cleaning the sensing device that controls defrosting, as shown below or in the picture at right.

Clearing an air-flow sensing tube. Heat pumps like the one below have a tube at the top of the unit that is connected to an air-pressure-activated switch. As long as the fan is freely sucking air in over the coils, pressure in the tube keeps the switch open. When ice build-up on the coils blocks the flow of air to the fan, pressure in the tube drops; the switch closes and turns on the defrost circuit. Spiderwebs or other foreign material may block the tube and keep the unit from switching to defrost. Unscrew the nut that holds the tube to the switch, thread a wire through the tube, attach a bit of cloth to the wire and pull the cloth through the tube. Refasten the tube to the switch.

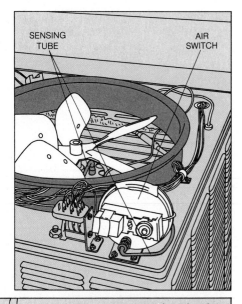

SENSING TUBE   AIR SWITCH

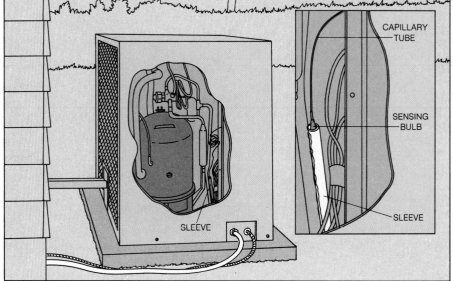

CAPILLARY TUBE

SENSING BULB

SLEEVE

SLEEVE

Cleaning a sensing bulb. Heat pumps like the model at right have a thermostat at the end of a thin copper tube that is either held by a clamp to the pipes along one edge of the coil or is located inside a sleeve attached to a refrigerant pipe. When the coil temperature drops below 32°, causing frost to collect, this thermostat switches on the defrost circuit. Corrosion may interfere with its sensitivity. Clean the bulb with sandpaper and the sleeve with a thin, spiral metal brush.

Troubleshooting Air Conditioners and Heat Pumps

Symptoms	Causes	Remedies
Air conditioner does not run	Defective cord or plug (room unit)	Replace cord and plug.
	Faulty thermostat (room unit)	Test thermostat; replace if necessary (page 121).
	Inoperative switch (room unit)	Test switch and replace if necessary (page 121).
Air conditioner operates continuously or goes on and off repeatedly	Faulty thermostat (room unit)	Test thermostat; replace if necessary (page 121).
	Thermostat improperly located (central unit)	Reposition thermostat (page 16).
	Sensor bulb out of position (room unit)	Reposition sensor bulb (page 121).
Air conditioner cools ineffectively	Dirty filter*	Replace disposable filter; wash reusable filters and coat with light oil.
	Clogged grilles and dirty blower or fan*	Loosen dirt with a stiff brush or putty knife and then clean up dirt with a vacuum cleaner.
	Blower drive belt too loose (central unit)	Tighten belt (page 37).
	Faulty thermostat (room unit)	Test thermostat; replace if necessary (page 121).
Frost on evaporator coil	Unit turned on when outdoor temperature is below 60°*	Do not operate when temperature is below 60°; severe ice build-up can break refrigerant lines.
	Dirty filter*	Replace disposable filter; wash reusable filter and coat with light oil.
	Bent coil fins*	Straighten fins with special fin comb available from refrigeration-supply dealers.
	Blower drive belt too loose (central unit)	Tighten belt (page 37).
	Faulty thermostat (room unit)	Test thermostat; replace if necessary (page 121).
	Coil clogged with dirt*	Clean coil with vacuum cleaner.
Water leaks into a room	Air conditioner tilted (room unit)	Adjust level of unit (page 102).
	Condensate drain hole plugged (room unit)	Clean drain with a coat hanger.
	Condensate box and pump clogged (central unit)	Clean box and pump (page 120).
Blower motor overheats	Motor needs lubrication (central unit)	Oil motor (pages 36 and 38).
	Drive belt too tight (central unit)	Adjust belt tension (page 37).
Excessive noise or vibration	Loose grilles or access panels*	Tighten screws or secure with tape.
	Thermostat sensor bulb touching coil (room unit)	Bend sensor bulb away from coil; replace rubber washer that secures sensor if worn.
	Incorrect drive belt tension (central unit)	Adjust tension (page 37).
	Drive pulleys misaligned (central unit)	Realign pulleys (page 37).
	Blower motor loose on mount (central unit)	Tighten mounting bolts (page 36).

*These causes and remedies apply to both room and central units.

Calculating Your Cooling and Heating Needs

How big a furnace—oil, gas or electric—do you need for your house? How big an air conditioner? The answers, whatever the size, condition or location of the house, can be derived from a few simple observations plus some simple arithmetic. First you must take into consideration where you live and the quality of your weatherproofing *(below)*: geography determines not only your primary heating or cooling needs but also the insulation and weatherproofing you should have; tighter weatherproofing can lower the cost both of buying and of running your new equipment.

The capacity of heating and cooling equipment is generally rated in British thermal units per hour, abbreviated as BTUH. (A single BTU is the amount of heat needed to raise the temperature of one pound of water by one degree.) For a gas or oil furnace, the BTUH rating is usually given as one of the basic facts of the unit; for an electric furnace, which is generally rated in watts, multiply the wattage by 3.4 to get the BTU equivalent; and for an air conditioner, multiply the tonnage by 12,000 for the same purpose.

The calculation of your own BTUH needs begins with the climatic maps opposite. The summer cooling map is zoned for typical maximum temperatures, the winter heating map for typical minimum temperatures and the effects of winter winds. In both maps, the zone numbers form the basis for BTUH calculations.

The numbers on the summer cooling map *(opposite, top)* indicate the area in square feet that can be cooled by 12,000 BTUH of equipment in a house with average weatherproofing. The BTUH requirements for a house with tighter or looser weatherproofing must be adjusted with a correction factor.

Determining BTUH for heating *(opposite, bottom)* is more complex. Once you find the base BTUH figure, and adjust it if necessary for a tight or loose house, addi-tional calculations give a BTUH figure for a gas or an oil heater. A final calculation gives the BTUH for a furnace used to heat water as well as space.

The tightness or looseness of your weatherproofing depends upon factors of construction and quality. By far the most important of these factors is the amount of insulating material in the walls, floors and ceilings; others have to do with the sealing of chinks and cracks. A tight house has vapor barriers, tight storm doors and windows, new weather stripping and caulking, and more insulation than the average for its climatic zone, as listed in the insulation chart below. An average house has vapor barriers, somewhat loose storm doors and windows, and the average insulation listed in the chart. A loose house—typically, one built before 1930 and not insulated since—has no storm doors or windows, no weather stripping, caulking or vapor barriers, and little or no insulation.

Determining Your Insulation

Meeting a variety of conditions. The zones on this map reflect insulation practices ranging from those of Alaska, where temperatures may drop to -50°, to those of Florida, where the thermometer may rise above 100°. Used with the chart at right below, these zones indicate the average insulation needed for both cooling and heating. In northern regions, minimum winter temperatures and wind-chill factors determine insulation needs: the insulation works to hold heat within the house. In the South, high summer temperatures determine the needs: insulation works to keep heat from flowing into the house.

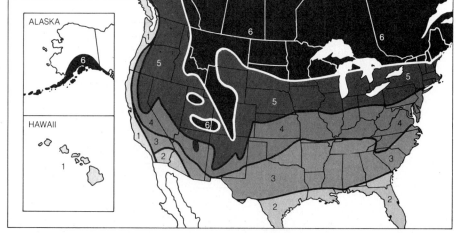

Average amounts of insulation material. The chart at right lists the actual R-values—a measure of a material's resistance to heat flow—of the insulation materials generally installed in well-built houses over the last 20 years. To use the chart, determine the R-value of the insulation in your home, and check it against your zone number on the map above to see if it is above or below average. If you live on the border between two zones, choose the zone with the higher value. Insulating material is installed only within certain ceilings, walls and floors: in a ceiling that is below a roof or an unheated attic; in an ex-terior wall; and in the floor if it is over an unheated basement or crawl space.

If you do not know the R-value of your insulation, check unfinished attics, basements or crawl spaces. If the insulation consists of batts or blankets, their R-values are printed on the surface. Inside walls there is no simple way to look for R-value markings. Poke around the outer edge of an electrical outlet box to pull out a sample of the insulation; take it to a building-supply dealer and ask him its R-value, assuming the thickness is 3½ inches, sufficient to fill the wall space.

Zone	Ceiling or roof	Exterior wall	Floor
1	R-9	R-9	R-7
2	R-11	R-9	R-9
3	R-13	R-11	R-9
4	R-19	R-11	R-11
5	R-19	R-13	R-11
6	R-22	R-13	R-13

Summer Cooling Zones

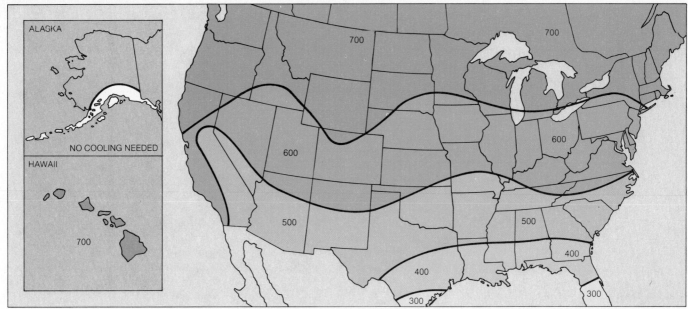

Calculating BTUH for cooling. Find your zone on the map. To determine the tons of refrigeration you need, divide the number of square feet that you wish to cool by your zone number; to convert the tonnage figure to BTUH, multiply it by 12,000. Thus, a house with a 500 zone number and 2,000 square feet of floor space needs 4 tons, or 48,000 BTUH, of refrigeration.

This figure is for a house with average weatherproofing (*opposite, top*). A house with tight weatherproofing will require less refrigeration; a house with loose weatherproofing will need more. If your house is tight, multiply the base figure by 0.85—giving, in this example, an actual requirement of 40,800 BTUH. For a loose house, multiply the base figure by 1.3 to get the actual requirement—62,400 BTUH.

Winter Heating Zones

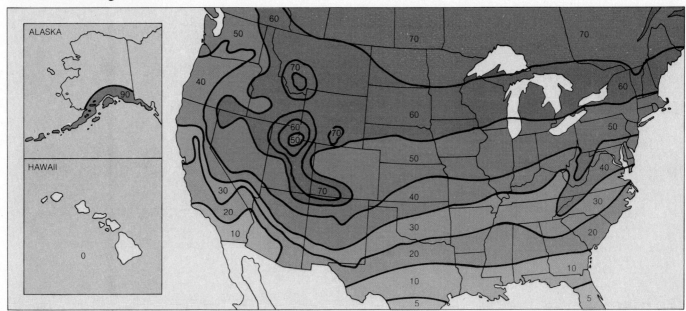

Calculating BTUH for heating. Find your zone on the map. Multiply your zone number by the square feet of floor space to be heated to find the base amount of BTUH. For a tight house, multiply this figure by 0.7; for a loose house, multiply it by 1.5. For example, an electrically heated house in Zone 30 with 2,000 square feet and average weatherproofing needs 60,000 BTUH of equipment. If tight, the house needs 42,000 BTUH; if loose, 90,000 BTUH. To calculate the requirements for a gas or oil furnace, you must raise the adjusted base capacity. For gas, multiply the adjusted base capacity by 1.25; for oil, by 1.3. (In the example above, the loose house requires a 112,500-BTUH gas furnace, or a 117,000-BTUH oil furnace.) If the heating system also heats the water supply, multiply the required capacity by 1.2. The loose house in the example would require a 108,000-BTUH electric furnace, a 135,000-BTUH gas furnace, a 140,400-BTUH oil furnace.

Picture Credits

The sources for the illustrations in this book are shown below. Credits for the pictures from left to right are separated by semicolons, from top to bottom by dashes.

Cover—Fred Maroon. 6—Fred Maroon. 8 through 13—Drawings by Dale Gustafson. 14 through 19—Drawings by Vantage Art, Inc. 20—Fred Maroon. 22 through 25—Drawings by Vantage Art, Inc. 26 through 32—Drawings by Peter McGinn. 34,35—Drawings by Nick Fasciano. 36 through 41—Drawings by Ray Skibinski. 42 through 47—Drawings by Whitman Studio, Inc. 48—Fred Maroon. 50 through 59—Drawings by Peter McGinn. 60,61—Drawings by Vicki Vebell. 62 through 65—Drawings by Mitchell Kuff. 66 through 69—Drawings by Ray Skibinski. 70,71—Drawings by Vantage Art, Inc. 72 through 77—Drawings by Edward Vebell. 78 through 85—Drawings by Vicki Vebell. 86 through 91—Drawings by Great, Inc. 92—Fred Maroon. 94 through 99—Drawings by John Sagan. 100,101—Drawings by Nick Fasciano. 102 through 107—Drawings by Gerry Gallagher. 108 through 119—Drawings by Adolph E. Brotman. 120,121—Drawings by Vicki Vebell. 122 through 125—Drawings by Gerry Gallagher.

The following persons also assisted in the making of this book. Marie Bradby, Charles Crocker, Lewis Diuguid, Joyce Doherty, Steven Forbis, William Garvey, Joseph B. Goodwin, Andy Leon Harney, Harvey Kahn, Anne Henehan Oman, Curtis Prendergast and Virginia Seippel helped with the writing and research. Fred Holz, Gerry Gallagher, Fred Collins, W. F. McWilliam, Susan Starkweather, Joan S. McGurren and Carol Summar prepared the sketches from which the final illustrations were drawn.

Acknowledgments

The index/glossary for this book was prepared by Mel Ingber. The editors also thank the following: John C. Adams, Public Relations Director, Fedders Corp., Edison, N.J.; Alvin Anderson, Service Dept. Mgr., Fasco Industries, Inc., Fayetteville, N.C.; William Axtman, American Boiler Manufacturers Assoc., Arlington, Va.; Joe Burkhardt, Jim Simpson, Accurate Air System, Hyattsville, Md.; Richard J. Costello, Mgr., Scott's Lumber & Garden Centers, Alexandria, Va.; John M. Davis, Thomas J. Fannon & Sons, Alexandria, Va.; John Deel, Dominion Electric Supply Company, Inc., Arlington, Va.; Beatriz de W. Coffin, Coffin & Coffin, Landscape Architects, Washington, D.C.; R. J. Denny, Assistant Director of Engineering, Air-Conditioning and Refrigeration Institute, Arlington, Va.; David M. Dolinsky, Mgr., Sandmeyer Oil Company, Cornwall Bridge, Conn.; Spencer Dormitzer, Beacon-Morris Corp., Boston; Ralph Eisenbeis, Hydronics Division, Burnham Corp., Lancaster, Pa.; David Essen, Mgr., Turbonics, Inc., Cleveland; Denis Fallon, Climate Control Specialist, Consumer Products Division, Fasco Industries, Inc., Fayetteville, N.C.; Gary Fike, Mgr., R. E. Michel Company, Inc.; Arlington, Va.; George Grena, Vienna, Va.; Jim Halleran, Annandale, Va.; Roger A. Hammer, Public Relations, Honeywell, Inc., Minneapolis; John A. Hill, Jr., President, Ultimate Engineering Corp., Natick, Mass.; Walter Hite, J. and H. Aitcheson, Inc., Wholesale Plumbing & Heating Supplies, Alexandria, Va.; Charles R. Holm, Residential Division, Honeywell, Inc., McLean, Va.; Edward F. Joyce, Capital Hydronic Supply Company, Falls Church, Va.; Robert A. Kaplan, Vice President & Director of Engineering, Wayne Home Equipment Company, Fort Wayne, Ind.; Morris Katz, Jerry Taylor, Acme Stove Company, Washington, D.C.; Kings Row Fireplace Shop, Springfield, Va.; Louis F. Kurtz, The Hydronics Institute, Berkeley Heights, N.J.; Bill Lehmberg, Goldkist Golden Flame Logs, Atlanta; William Lewis, McGraw-Edison Company, Air Comfort Division, Albion, Mich.; Donald C. McClurg, Associate Director, Home Ventilating Institute, Chicago; Bill Milon, Staff Engineer, Bachrach Instrument Company, Pittsburgh; Samuel Pritchett, Walter D. Sutton, Inter Technology Solar Corp., Warrenton, Va.; John Purcell, Top Notch Laminates, Rockville, Md.; Eliot Sepinuck, Beacon-Morris Corp., Boston; Paul Smith, Empire Stove Company, Belleville, Ill.; Glenn H. Spoerl, Group Public Relations Mgr., Sears Roebuck, Chicago; Thola Theilhaber, Corporate Conservation Engineer, Raytheon Company, Lexington, Mass.; David Von Gillern, ITT Fluid Handling Division, Bell & Gossett, Morton Grove, Ill.; Wallace-Eannace Associates, Inc., Plainview, N.Y.; Clint Walls, Slant/Fin Corp., Greenvale, N.Y.; Bert J. Watling, President, The Carlin Company, Wethersfield, Conn.; Burt Weller, Staff Engineer, National Oil Jobbers Council, Washington, D.C.; Ben Zimmer, Underwriters' Laboratories, Northbrook, Ill.

Index/Glossary

Printed in U.S.A.